The Hidden Frontier

Ecology and Ethnicity in an Alpine Valley

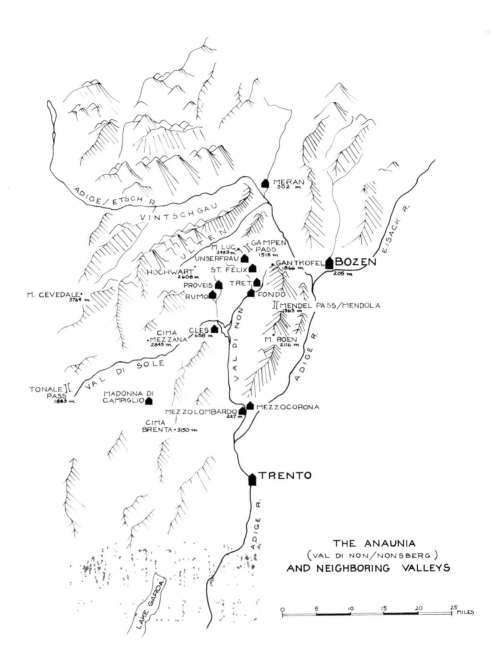

ADIGE / ETSCH R.

VINTSCHGAU

MERAN
302 m

EISACK R.

ULTEN

M. LUC.
2483 m

GAMPEN
PASS
1518 m

UNSERFRAU

GANTKOFEL
1866 m

BOZEN
205 m

HOCHWART
2608 m

ST. FELIX

PROVEIS

TRET

RUMO

FONDO

MENDEL PASS / MENDOLA
1363 m

M. CEVEDALE
3764 m

VAL DI NON

CIMA
MEZZANA
2845 m

CLES
658 m

M. ROEN
2116 m

VAL DI SOLE

ADIGE R.

TONALE
PASS
1883 m

MADONNA DI
CAMPIGLIO

MEZZOLOMBARDO
227 m

MEZZOCORONA

CIMA
BRENTA • 3150 m

TRENTO

ADIGE R.

THE ANAUNIA
(VAL DI NON / NONSBERG)
AND NEIGHBORING VALLEYS

LAKE GARDA

0   5   10   15   20   25
MILES

The
Hidden Frontier

Ecology and Ethnicity
in an Alpine Valley

John W. Cole
Department of Anthropology
University of Massachusetts
Amherst, Massachusetts

Eric R. Wolf
Department of Anthropology
Herbert H. Lehman College
Bronx, New York

ACADEMIC PRESS   New York and London
A Subsidiary of Harcourt Brace Jovanovich, Publishers

ACADEMIC PRESS, INC.
111 Fifth Avenue, New York, New York 10003

United Kingdom Edition published by
ACADEMIC PRESS, INC. (LONDON) LTD.
24/28 Oval Road, London NW1

Library of Congress Cataloging in Publication Data

Cole, John W
    The hidden frontier.

    (Studies in social discontinuity)
    Bibliography: p.
    1.   San Felice, Italy.   2.   Tret, Italy.   3.   Anthrop
-geography—Trentino-Alto Adige, Italy.   4.   Germans in
Bolzeno (Province)   I.   Wolf, Eric Robert, Date
joint author.   II.   Title.   III.   Series.
[DNLM: 1.   Anthropology, Cultural—Switzerland.
GN585.S9 C689h 1973]
DG975.S1857C64   301.29'45'38   72-13621
ISBN 0–12–785132–1

For Ellan, from John
For Katia, from Eric

STUDIES IN SOCIAL DISCONTINUITY

Under The Consulting Editorship of:

CHARLES TILLY
University of Michigan

EDWARD SHORTER
University of Toronto

William A. Christian, Jr. Person and God in a Spanish Valley

Joel Samaha. Law and Order in Historical Perspective: The Case of Elizabethan Essex

John W. Cole and Eric R. Wolf. The Hidden Frontier: Ecology and Ethnicity in an Alpine Valley

In preparation

Immanuel Wallerstein. The Modern World-System: Capitalist Agriculture and the Origins of the European World-Economy in the Sixteenth Century

D. E. H. Russell. Rebellion, Revolution, and Armed Force

John R. Gillis. Youth and History

Kristian Hvidt. Flight to America

Contents

Preface

In opening this book our readers will discover that both in the presentation and analysis of our materials we have followed a scheme somewhat at variance with the ways usually followed by anthropologists. We strongly believe that the study of small populations which form components of complex societies must take account of that complexity before the interpretation of what happens "on the ground" can become meaningful. Thus we believe that anthropology cannot do without history, for it is only through an anthropologically informed historical account of the genesis and development of the forces impinging upon our social and cultural microcosms that we can arrive at an adequate assessment of the ways in which these forces act upon each other in the present. We thus begin with history: first with an account of the forging of Tyrolese identity, followed by a discussion of how that identity was subjected to the contradictory pull of rival nationalist movements. We then move, in Chapter IV, to an historical account of the economic forces at work in shaping the life of the mountain peasantry. In Chapter V we attempt to state what we know of past happenings in the Anaunia proper, in order to sketch the historical background of the two communities that we studied in the field. Chapters VI and VII then deal with the problems of wresting a living from a forbidding environment and a marginal economy. In Chapter VIII we examine the inheritance process, those mechanisms which operate ecologically and socially to divide available resources among possible claimants. In Chapter IX we present our assessment of the changes that have affected both Tret and St. Felix during the economic upswing of the past two decades. This, in turn, leads us to an explication of the bonds of kinship, friendship, and neighborliness characteristic of the two communities. We return, in the last chapter, to a discussion, this time on the local level, of the ways in which contradictory political commitments pull people in opposite directions, despite their very similar ways of meeting the problems of survival and existence in a mountainous environment.

Acknowledgments

In the course of this study, which attempts to combine the methods of anthropological field work with the perspectives of history and political economy we have relied heavily on the assistance and inspiration granted us by numerous friends and colleagues in many different fields. Some of them provided us with valuable background information; others were generous with introductions. Among these we would like to thank especially Frank and Julie Fata, Emanuel Fenz, Prof. Raymond Firth, Dr. E. K. Francis, Prof. Luis Hechenbleikner, Dr. Franz Huter, Dr. Hans Kramer, John Martin, Dr. Norbert Mumelter, Prof. Tullio Tentori, Dr. Leo von Pretz, Dr. Guido G. Weigend, and Prof. Anton Zelger. We are heavily indebted to our friends in St. Felix and Tret, notably Alois Ausserer ("Pinterluis"), Livio Bertagnolli and family, Adele Donà, Martin Geiser ("Schmied") and family, Stephanie Geiser, Curate Bartolomäus Hillebrand, Josef Kofler ("Ludwigsepp"), and Emanuele Maninfior ("Mani"). We are thankful to the many archivists who allowed us to examine the documents in their charge, notably P. Tito of the Capuchin monastery in Cles and Don Alessandro Sartori. We acknowledge with pleasure the special stimulus received from fellow participants in two fieldwork conferences, sponsored by the Mediterranean Study Groups of the University of Michigan and of the London School of Economics–University of Kent at Aix-en-Provence in 1966 and at Canterbury in 1967. Raymond Grew, Roger McConochie, Donald Pitkin, Joyce Riegelhaupt, Roland Sarti, and H. Martin Wobst aided our work with criticism and suggestions. Sydel Silverman did much to bring the project to fruition; she queried our ideas and castigated our style and easy ways with punctuation. We are most grateful. We would also like to thank Foto Battisti in Fondo for aid with our attempts at photography, to Ruth Fair who typed our manuscript, and to Judy Hammond who drew our maps.

Our work would not have been possible without the financial support so generously granted us by various institutions. Eric R. Wolf's research in 1960–1961 was supported by a grant from the John Simon Guggenheim Foundation, and by a grant-in-aid from the Lichtstern Fund of the University of Chicago. His work in the area in 1962 was underwritten by funds from the Horace Rackham School of Graduate Studies of the University of Michigan and the Social Science Research Council. Subsequent visits to Tret and St. Felix were made possible by his tenure of a Career Development Award from the National Institute of Health (5 K02 MH 2543405). John W. Cole's initial stay during 1965–1967 had the support of a grant from the National Science Foundation, with supplementary grants during that period and again in 1969 offered by the Project for the Study of Social Networks in the Mediterranean Area of the University of Michigan and by Wayne State University. We are most thankful to all these institutions for their support, and hope that the present work in some small measure justifies the trust which they put in us during a decade of fieldwork and related research.

The Hidden Frontier

Ecology and Ethnicity in an Alpine Valley

CHAPTER I

The Inquiry

In this book we deal with two small villages located on the high alpine rim of northern Italy—two microcosms caught up in the play of forces larger and more powerful than themselves. It is our purpose to describe and explain this interplay between microcosm and macrocosm, between peasant settlement and an expanding market, between community and more inclusive polity. The two villages lie next to each other, but they are separated by provincial boundaries: German-speaking St. Felix today forms part of the Italian Province of Bozen or Bolzano, officially designated as Alto Adige or Tiroler Etschland; Romance-speaking Tret, its neighbor, belongs to the Province of Trento or Trentino. Both provinces formed, before World War I, an integral part of the Austrian *Land* of the Tyrol, and were thus component parts of the large Austro-Hungarian Empire (see Figure 1). Province Bozen then had a population of 285,000, of whom fewer than 10,000 spoke Ladinsh, a Rhateo-Romansh language, and another group of fewer than 10,000 members spoke Italian; the rest were German speakers (Fiebiger 1959: 17; Leidlmair 1958: 39). Province Trento, in comparison, counted 387,000 inhabitants; most of them spoke Italian. Ladinsh speakers and German speakers there numbered fewer than 14,000

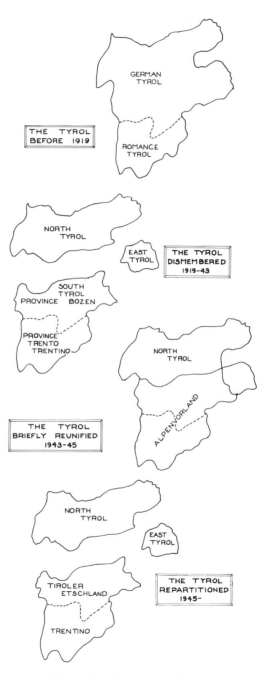

Figure 1. The Tyrol in history.

(Dörrenhaus 1959: 91; for population figures see Appendix 1 of this book).

After World War I both provinces were transferred to Italy and—with the exception of a brief interlude during World War II—have formed parts of the Italian state ever since. The political settlement at the end of World War I, in dividing the German-speaking Province of Bozen/Bolzano, or the South Tyrol, from the remainder of the Tyrol, thus also granted the Italian state political jurisdiction over more than 200,000 German speakers, who henceforth constituted an ethnic minority within Italy; the inhabitants of predominantly German-speaking St. Felix belong to that minority. In contrast, the largely Romance-speaking Province of Trento was claimed by the victorious Italian state as part of Italian territory liberated from Austrian domination. For the Italians, the Trentino was *terra irredenta*, "unredeemed land," and the Romance speakers of Tret, consequently, were incorporated into the Italian nation.

Yet these two villages, different in ethnic identity and often at loggerheads in politics, live side by side and share very similar modes of adaptation to a common mountainous environment. We are interested in the commonalities that unite them and in the social, cultural, and political oppositions that divide them. We hope, in the course of an historical and anthropological analysis of St. Felix and Tret, to come to understand the forces that have shaped them and to better understand the larger processes of economic and political involvement at work in the wider enveloping societies.

We know that a study of small populations will not reveal all there is to know about the total societies in which they are embedded, and we are similarly aware that the study of total societies will not in and of itself provide grounds for predicting how small populations react to more wide-ranging systemic processes. The total society is not a community writ larger, nor is a small settlement in the mountains a replica of a larger whole in miniature. Nevertheless, we are convinced of the utility of an approach that sees the *relation* between village and nation as problematic, and hence as a source of potential insights. That relation is for us not simple and mechanical; rather, we understand it as complex and dialectical. It is dialectical because village and total society exist in opposition and often in contradiction. It is dialectical because these two units in opposition interpenetrate each other and act upon one another in social and cultural interchanges. It is dialectical, finally, in that this interaction generates an ongoing transformation over time, which subjects the narrower unit to ever more comprehensive processes of integration, or synthesis. As anthropologists, we want to focus on the transformations worked upon small populations; but we also want to know something of the way in which what happens at the village level supports or stands in contrast to the character and direction of movement in the larger system. We are interested in the transformations of local ecological patterns and political alignments in relation to the promptings of market and nation-building.

Every study has a history and develops its own characteristic problems and solutions. In this chapter we want to address our-selves to the considerations that brought us to the high alpine villages of St. Felix and Tret in the upper reaches of the valley of Anaunia, and to detail the manner in which the study was carried out.

This inquiry had its formal beginnings in Eric Wolf's applica-tion to the John Simon Guggenheim Foundation for financial sup-port in 1957—its informal origins extend further into the past. Wolf had visited the South Tyrol as a child, in the summer of 1934, for a stay in the Val Gardena, or Gröden Tal, a popular tourist center. It was an eventful summer outside the valley: Ernst Roehm and his left-wing National Socialists were slaughtered by Hitler in June, and in July, Chancellor Dollfuss of Austria was assassinated in Vienna by the Nazi underground. The Gardena, by contrast, seemed idyllic in its peacefulness; yet one could feel undercurrents of hostility be-tween the inhabitants of the valley and the officialdom of a Fascist Italy bent on their forcible acculturation. Having grown up first in Vienna and then among the Germans and Czechs of the embittered Sudeten frontier of Czechoslovakia, even a boy could not help but become sensitized to the conflicts of ethnicity and nationalist loyal-ties left unresolved by the collapse of the Habsburg Empire; long before he became an anthropologist, Wolf was led to ask of himself and others why ethnic and nationalist loyalties so often crosscut allegiances of class or formal citizenship.

Wolf saw the area again, at the end of World War II, as a member of the United States' mountain troops, and once more wit-nessed the hostile encounter between German-speaking Tyrolese and Italians. The anthropological question of the why and where-fore of ethnic boundaries and loyalties arose once more, still with-out satisfying answers. While his later interests in graduate studies shifted to Southeast Asia and Latin America, Wolf did maintain a continuing concern both with the narrower problems of the divided Tyrol and with the larger problems posed by the existence of ethnic conflicts and accommodations. Thus, in 1947–48, he interviewed

Tyrolese in New York for Columbia University's Research in Contemporary Cultures, directed by Ruth Benedict, and his first field work in Mexico in 1951 was prompted by an interest in the genesis of Mexican national identity. Hence, in the late 1950s, it seemed worthwhile to initiate a study aimed specifically at gaining a clearer understanding of ethnic contrasts. Several European regions suggested themselves—for example, Schleswig-Holstein or Cyprus. Since either study would have meant learning Danish, Turkish, or Greek, Wolf decided on the South Tyrol in the Italian Alps: he was familiar with the area, spoke German as a mother tongue, and had learned some Italian in the course of the war. He selected for intensive attention the twin villages of St. Felix and Tret. St. Felix is German speaking, whereas the people of Tret speak a variant of Romansh. The two village centers are located no more than a half-hour's walk from each other, one mile apart as the crow flies, on a mountain plateau south of Meran and west of Bozen. This mountain plateau is called the Nonsberg by the German-speaking Tyrolese, and the Val di Non by Italians. To avoid the ethnic connotations that these names suggest, we shall refer to it hereafter by its Roman and Medieval name, the Anaunia.

The South Tyrol was selected as the general area of study, and St. Felix and Tret chosen as the particular villages to be studied, to allow us to investigate ethnic coexistence or conflict with a minimum of additional linguistic preparation. On the face of it, this criterion proved to be the product of an undue optimism. Although Viennese will readily tell you that they experience no difficulty in understanding Tyrolese, their claim is not borne out in practice. Tyrolese, in turn, experience a language gap between their own dialect and standard German, and veterans of the Galician campaign of World War I have told us that they could understand Yiddish, equally a peripheral dialect of German, more easily than the speech of other Austrian soldiers. In Galicia, they averred, they encountered "lauter Tiroler," many Tyrolese; however, these Tyrolese wore long, black coats, sidelocks, and black hats. Nor did familiarity with Italian ease communication in Nónes, the Romance language spoken in the Anaunia. Nónes belongs to the Romansh family of languages, and Wolf's knowledge of French and Spanish often proved more useful than Italian.

There were other unforeseen problems, both practical and methodological. Finding sufficient firewood for cooking and heating in villages not provided with facilities for lodging outsiders proved time consuming. Working in an area with a long tradition of record keeping required a shift from more synchronically oriented fieldwork to a more workmanlike concern with local and regional history. Most obvious of all difficulties, however, was the challenge of a field situation that proved too close, both psychologically and culturally, to Wolf's Austrian background. This was least obvious in the collection of genealogical data and family histories, and most obvious in the recording of information on political history, especially where past experiences with National Socialism and Fascism still obtruded into the ethnographic present. After the initial year of fieldwork, 1960–61, and a return visit in the summer of 1962, Wolf, therefore, sought to counteract his self-perceived lack of distance. Ideally this could best be done by inviting a restudy of the two villages by another anthropologist who could add perspective. The work was thus taken up by John W. Cole, then a graduate student at the University of Michigan, who carried on independent field work in the same area for eighteen months, from September 1965 to March 1967.

Cole complemented Wolf's inquiry in several ways. He brought to the study an interest in cultural ecology, as well as a measure of previous acquaintance with the culture and language of the area, gained between 1957–60 while serving with the United States Army in Germany. Frequent trips into the Alps provided him with welcome relief from the regimentation of army life and from the insularity of the American community around the army base. In beginning his own field work he could draw on Wolf's data on social organization as a basis for his own study of village ecology and economy. The timing of his fieldwork also proved auspicious, since he came to live in the villages at a time when agricultural machinery made its first appearance and the migration to take up work in the adjacent lowlands or in Germany had begun in earnest. These events marked a shift toward both a greater commitment to commercialized agriculture and nonpeasant sources of subsistence. Cole's restudy thus allowed us to witness a major change in response to external factors, and offered some insight into the consequences of a growth in economic well-being for villages poorer and more isolated at the start of the decade. Wolf again visited the area in the summers of 1965 and 1968, and both authors returned in the summer of 1969.

The selection of St. Felix and Tret as field sites requires still further explanation. The ethnic frontier between German and Romance speakers is divisible into many sectors; any of these would have furnished an equally propitious—and certainly more comfortable—location for study. Some readers may find it ironic that we selected one of the least known and most inaccessible of these. We could have returned, for example, to Gröden, with its lush, upland meadows, its villages of wood carvers, and its tourist hotels. We could have settled in the Unterland, the vineyard region in the Etsch River valley below Bozen. We could even have studied ethnic conflicts and symbiosis in the twin cities of the area, Bozen and Meran. Instead, we selected villages in a little-known valley: we would comment jokingly about the villages as examples of Danubian II, and came to call the Val di Non the "valley of nothing." But we purposely made this selection. First, St. Felix and Tret provided us with an opportunity to study ethnic interrelations that were age-old and not merely the product of recent exacerbations of ethnicity due to political conflict. Second, we wanted a simple field situation in which we could get to know all the relevant actors and participate as fully as possible in their various activities. This meant that we wanted to avoid cities and towns. Urban anthropology, while an important and growing branch of anthropology, had not yet developed the research tactics and strategy that would have justified an extended stay in Bozen or Meran. Third, we wanted to avoid areas where cultural and political conflicts were so extreme that we could have worked only with one side, to the exclusion of the other. Fourth, we did not want to deal with a situation in which the annual influx of large numbers of tourists would blur our research focus—though we are fully aware of the importance of tourism as a cultural phenomenon demanding study in its own right. At any rate, in St. Felix and Tret there were no tourists—only a handful of summer guests with ties in the villages and in no position to reshape or drastically restructure the ongoing way of life.

In choosing St. Felix and Tret we thus followed the anthropologist's traditional penchant for the study of relatively isolated and "unspoiled" situations, but we make no apologies for the choice. We believe, in fact, that, if anything, the study of poor, peripheral communities has been rather slighted in the study of complex societies; even in well-studied Mexico, there are few studies of poor, isolated Mije or Otomí-speaking populations. Tret and St. Felix are certainly not statistically representative of the entire gamut of South Tyrolese and Trentino communities. However, they are characteristic of the social and ecological world of mountain peasantry in both the South Tyrol and the Trentino, the world of villages in which more than 50% of the economically engaged population carries on agriculture and livestock-raising (Leidlmair 1958:

108). Such villages abound in Schnals, Martell, Ulten, Sarntal, and in the periphery of the Valley of the Puster in the South Tyrol; in the Giudicarie, Val di Sole, Fiemme, or Fassa, within the Trentino. St. Felix and Tret are, of course, not representative of the cultivator–wageworker villages of the Vintschgau, nor of artisan and craft-producing villages near Meran and Bozen; nor do they reveal anything of the life of fruit growers and vintners along the Etsch. They are, nevertheless, diagnostic in exhibiting, in stark form, social and cultural processes that remain more muted elsewhere, since their close coexistence for more than seven hundred years permits insights into their ecological and ideological differences in particularly profiled form.

TWO VILLAGES

Some of the ecological and ideological differences are apparent in their external structures. Tret is a nucleated village of thirty-seven inhabited dwellings. Multiple farm structures are the most common type; the thirty-seven structures contain a total of sixty inhabited apartments. At the center of the village is its largest building, the *setti-communi*, which contains several apartments and barns, as well as one of the two village stores. Three other structures, including the second store, adjoin the *setti-communi*, forming what the villagers jokingly call their *piazza*, or village square. Fields in Tret characteristically lie outside the settlement area, extending for some distance in every direction; holdings are fragmented—the plots belonging to a household may be widely dispersed. St. Felix, in contrast, is characterized by a predominantly scattered pattern of settlement. Its core consists only of the church, the tavern (which also houses a store), and four other buildings; all other houses and barns lie scattered over the landscape, each surrounded by a solid block of fields and meadows separating it from its nearest neighbor. St. Felix possesses sixty-one such buildings, comprising a total of seventy-four apartments. Each building, with its attached fields, meadows, and pastures, bears a name that endures over the generations: its occupants characteristically bear that name and are called by it, rather than by their legal surname. In Tret, houses also bear names, usually those of some past occupant, but people are referred to by nicknames or the name of the stock to which they belong, in order to differentiate, for instance, two Bepo Bertagnollis as Bepo

"Tret is a nucleated village of thirty-seven inhabited dwellings." At the beginning of spring, fields and meadows will receive their cover of manure.

In St. Felix, in contrast, "houses and barns lie scattered over the landscape."

Cru and Bepo Franzett. The difference in settlement patterns and in the layout of fields in the two villages is illustrative of a more general contrast between the rural Trentino, which tends toward a pattern of nucleated settlements and dispersed holdings, and the rural South Tyrol, with its characteristic pattern of isolated homesteads and compact estates (see Dörrenhaus 1959).

The two villages also illustrate a more general political contrast between the Romance-speaking Trentino and the German-speaking South Tyrol. The rural community in the Trentino, as in most of Italy, characteristically forms part of a political entity governed from the seat of the municipal government, which is located in an urban center. Rural settlements do not administer their own affairs but send delegates to the dominant town. Thus, Tret is a *frazione*, or *ward*, of the commune of Fondo; it has a *capo frazione*, a ward headman who is under orders to the town government; and it sends two delegates, the *capo* and one other, to represent its interests in the meetings of the communal council in Fondo. In contrast, St. Felix—like other South Tyrolese rural communities—is a self-governing commune, with its own mayor (*Bürgermeister*), its own elected communal council of twenty heads of households, its own committee regulating access to forest and pasture land, and a committee to permit or prevent transfers of land.

The contrast between the dependence of the Italian rural settlement on an urban center and the autonomy of the Tyrolese rural settlement is illustrated still further in St. Felix and Tret in ecclesiastical matters: St. Felix has its own priest to whom the community assigns a residence and farm lands for subsistence (*Widum*), rendering him comparable in the operation of a holding to other owners of homesteads. It also possesses a plethora of religious associations (see Chapter XI). In contrast, the priest serving Tret lives in Fondo and alternates his services to the *frazione* with attendance at mass in nearby Dovena.

The political and ecclesiastical contrasts have larger social implications as well. The Italian, and thus Trentine, cultivator, the *contadino*, belongs to a *contado*, the rural orbit of a city. It is the city that is regarded as the seat of civilization and urbanity; the *contadino* is defined not merely as a second-rate citizen in a polity where urban dwellers take precedence, but also as an individual lower on the social scale, lacking in the civilized graces. The Tyrolese peasant, on the other hand, the *Bauer*, is not merely the owner of a homestead, but as such, holds political rights in his community and in a politically defined peasant estate within the Tyrolese assembly.

FIELDWORK

In fixing our attention on St. Felix and Tret, we did not choose communities representative of the wealthy and often opulent life of the lowland valleys, but poor and isolated settlements with few specialties and little chance to share in the richness of cultural forms so characteristic of the more prosperous Tyrolese peasantry. Poverty and isolation have also limited the number and kinds of social honors the communities can bestow upon their members. In St. Felix, a man may be elected mayor or communal councilman; in Tret, a man may be chosen ward headman or representative to the town council. Beyond these formal rewards, most tokens of recognition are informal, granted by fellow villagers to the successful manager of a holding; to the individual with a cool head who can give reasonable advice; to the woman who raises a garden of beautiful vegetables; to the man who is continually successful at cards. Being informal, these tokens of esteem—and conversely, their withdrawal—are based largely on individual performance. We soon became aware of the great range of idiosyncratic behavior among our informants and of the value granted, in gossip or social interaction, to the individual characteristics of particular men and women.

The two villages are poor, and diet is meager to the point of insufficiency. It consists of dumplings, potatoes, sauerkraut, and cheap wine, supplemented by smoked bacon on festive occasions or during periods of intense physical activity. Also lacking are most of the elaborations of art and ritual that make visits to other parts of the Tyrol, both North and South, rewarding to the urban visitor. Here there is no separate folk costume, nor idiosyncratic forms of music and song. Only one family does a little wood carving, and the use of decoration in painting interiors and exteriors is minimal, in contrast to the elaborate murals that often decorate Tyrolese houses elsewhere. The comfortable wood paneling and beautiful tile stoves of other Tyrolese houses are also absent. Religious ritual and art is not elaborate. What is lacking in other art forms, however, is made up in verbal play—and here our skills as linguists were not always up to the exigencies of the field situation. Though Wolf spoke

German as his mother tongue, he often could not follow joking except in translation or after careful explanation; he acquired only a smattering of Nónes, in which the Trettners communicate as in their very own "secret" language. Cole acquired a good speaking and reading kr.owledge of German, and a passable acquaintance with standard Italian. We are not, after this fieldwork, too sanguine about our capacity to claim firm command of the informal content of communication among the people we studied.

Though we have talked to all the people in both communities and participated in many different events during more than three combined years of fieldwork, we shall be quite content if we can explicate the more formal and structural aspects of village life. We will be exceedingly pleased if our work prompts other investigators, with greater skills in linguistics and more psychologically oriented inquiries, to carry out further work on the more informal aspects of communication in these two villages.

Although we hoped to divide equally our residence between St. Felix and Tret, both of us were forced to take up lodging in Tret, since no housing was available in St. Felix. Wolf, in 1960, found lodging for himself and his family on the second story of a house that had just become vacant upon the death of the owner's father. Cole, in 1965, occupied, with his family, a basement apartment, also recently vacant due to the death of a former tenant. On summer visits we stayed in the small inn in Tret, but in the summer of 1969, Cole and his family rented a newly built cabin above the village. During the first month of his year in the field, Wolf briefly had use of a car; after that he walked or took buses. In all later fieldwork we had access to a car, and could offer rides to villagers on their way to or from the market, or from the doctor in town. Yet, we did a great deal of hiking and walking over the extensive mountainside, inevitably becoming members of an information network that required everyone, German and Romance alike, to report to whomever he met the identities and whereabouts of all persons previously encountered on the trail. Also, like the villagers, we learned to walk through stands of forest with our eyes glued to the ground in search of potential firewood.

The people of Tret and St. Felix greeted our entry into the villages with predictable curiosity and a measure of reserve, but without much self-consciousness. There were rumors at the outset that Wolf's real reason for seeking a residence in the Upper Anaunia was its reputation for invigorating air: he had perhaps had a mental breakdown for which mountain air could have a salutary effect.

There did not seem to be anything else that could commend two villages, otherwise avoided by the tourist trade, to a visitor from America. At the same time, an interest in local history and customs seemed to be the sort of idiosyncrasy expected from a professor. Cole's entry was in turn eased by Wolf's previous stays, and his explanation that he was preparing a dissertation for an American university seemed sufficient for all concerned. Unexpectedly, perhaps, our acceptance in the villages was aided by the presence of older men who had once migrated to the United States and then returned; conversing with us with their few phrases of English validated their experience abroad in the eyes of other villagers, and we were able to rely on a few of these returnees for both advice and information.

We both were, during our first residence in the villages, accompanied by our families. Our wives, in charge of small children, were necessarily much tied to their own households, especially during the long and bitter winter months. This precluded for them, for instance, a role such as Katia Wolf had filled in previous field work in Puerto Rico, when she had been able to collect extensive data on the subculture of women and on the socialization of children. On the other hand, we also had children who attended school in Tret, and from whom we learned a great deal about both schoolmates and their parents. Wolf's six-year-old son, David, also became a kind of apprentice to the village smith, and in this capacity mastered much of the locally available metallurgical techniques. Both Cole's daughter, Sherry, and Wolf's son, David, developed a wide circle of acquaintances, and often visited village homes long before the authors. Subsequent summers we spent alone in the village supplemented our knowledge of the life patterns of unmarried males and females, as did Cole's participation in card games and Wolf's interest in skiing. The smith of Tret provided his own model of where we fitted into the local scheme of things. For him people fall into three categories: those who work with their hands, like peasants or artisans; those who work with their feet, like policemen and foresters; and those who work with their heads, like priests, teachers, and ourselves.

Due to the small size of the two villages, we in due course were able to come to know everybody and to participate in most structured public events. We watched people set out for work in the early dawn and talked to them during work breaks. We accompanied them to the fields when they spread manure on the melting snows of spring, and we walked with their herds in the annual movement to and from the high pastures. We took part in the critical task of bringing in the hay before the onset of rains ruined the hay crop, and we were invited to the festive cookouts of *polenta* (a maize porridge), sausages, and red wine held in the mountain meadows during haying season. We helped to gather mushrooms and wild flowers, and we were invited in to taste a particularly successful *grappa* brewed from eleven different wild plants, or to try a side of smoked bacon newly removed from the soot-covered chimney. Like other villagers, we shopped at the local stores and discussed the price and quality of produce. We also attended the regular markets in Fondo, and visited the cattle markets in Lana, Malé, Cles, and Bozen. We took part in the ceremonies surrounding a marriage; we attended funerals; and we walked in ritual processions, such as the Easter procession of 1969, which took place in a howling snow storm

Much of our information came to us in unstructured and informal interviews. We obtained a great deal of this kind of data in bars and taverns where men gathered to drink and talk. We engaged people in conversation whenever we could—we would talk to passengers on the bus or awaiting its arrival; we would wait for church on Sundays and talk to the assembling faithful; we would chat with men returning from a hunt, or with those who accompanied the horse-drawn plow used to clear village streets after a heavy snow fall. We took photographs of various events and persons, and gave copies to people who appeared in them, using the opportunity to discuss other features of interest to us. Wolf on occasion played the accordion at dances, and Cole spent a great deal of time playing cards—the favorite local pastime. His skill at card games varied still further the kinds of participant situations to which we had access. In card playing, old and young men interacted with ease: card playing was thus one of the few occasions when we could break out of the social definitions imposed on us by our roles as husbands and fathers.

We also did a considerable amount of formal interviewing. Cole especially sought systematic answers to a questionnaire (see Appendix 2). He interviewed members of sixty-one households, drawn in equal numbers from both villages. The interviews ranged in length from two hours to the longest, twenty-five hours, carried out on five different occasions. The average interview consisted of about two sessions of four hours each. About thirty villagers whom we interviewed became our friends and best informants. While we have striven to retain something of the color and warmth with

"Playing cards—the favorite local pastime."

which our informants responded to our queries and reported their own life experiences, we have forced ourselves to be quite rigorous in omitting all information that could in any way threaten the anonymity of the persons offering opinions and accounts, and that could identify them to their own detriment. We are aware that this will detract from the colorfulness of our account, but we feel most strongly that certain disclosures can infringe on an individual's right to privacy. We have not, however, disguised the names of the villages from which our data are drawn. In our account we shall aim at general statements, offered as characterizations of the two populations at large, rather than description and analysis of the action patterns and verbal reports of particular individuals.

It is our impression that our friends responded most easily to questions formulated around ecological concerns and most warily to queries that explored past political commitments. For instance, we still do not know all the names of the Trettners who participated in a partisan assault on the German command post in Unser Frau in the waning months of World War II, nor do we know the identity of all the Felixers who opted for removal to Germany. We have become convinced that in both communities the political movements of the period before 1939 and the war thereafter created conflicts so disruptive of social relations within and between the villages that most villagers are quite willing to forget the conflicts productive of such social tensions. Only an occasional dramatic event, notably the capture of Adolf Eichmann by the Israeli secret service and the acts of terrorism carried out by activist groups in favor of South Tyrolese reunion with Austria during 1961, would set villagers talking about political commitments and happenings of the past. Inevitably, because we worked at different times and with different emphases, we came to share some informants. At the same time each of us developed his own circle of friends and acquaintances; and one of the delights of repeated visits consisted in testing each other's impressions of the people we had come to know well separately, and to make the acquaintance of people who had become the friends of the other.

In addition to interpersonal interviews, we also worked to some extent with archival records. Wolf consulted the church records in St. Felix and Tret and the archives in Fondo, Unser Frau, and Trento. He also worked in the holdings of the Ferdinandeum and the University of Innsbruck. Cole consulted the registry offices in Fondo, Meran, and Cles, as well as the library of the Capuchin monastery in Cles. Both of us made extensive use of the holdings in the library of the Südtiroler Kulturinstitut in Bozen.

As we survey the data, we are reasonably certain of our ability to construct an interpretative model faithful to the events we have witnessed and the information we have collected both from informants and written records. We are also aware that our knowledge is by no means complete. We know more about some people than about others, more about Tret than about St. Felix, more about public events than about the private aspects of life, more about the lives of adults than about the lives of children, more about men than about women, more about ecology than about religion. We have better records on day-to-day interaction from Tret than from St. Felix, but we obtained better formal interviews in St. Felix than in Tret. Archival records also proved to be more complete in St. Felix than in Tret. We know that we have only partial answers to some of the questions that sent us into the field: in general, we have favored discussion of those topics on which we were able to gather reasonably complete information, rather than attempt to present a "well-rounded" study of a culture and society based on information that is highly variable in reliability.

We are well aware of the complex politics of the area, and we have tried sincerely to be dispassionate about the issues and reactions involved. We are both conscious of our personal sympathies for Tyrolese claims to continued cultural identity, though we refrain from judgments as to where the political boundary between Austria and Italy is now drawn, or whether it should be redrawn in the future. This stance has created some difficulties for us, not the least of which is posed by the use of various names for the area and its populations.

THE POLITICS OF PLACE NAMES

All place names carry implicit and explicit political connotations. Thus, the designation "Tyrol" goes back to a time when the now Austrian *Länder* of Tirol and Osttirol formed a common political province (within the Habsburg Empire) with the Romance-speaking Trentino and the predominantly German-speaking, but now politically Italian, Province of Alto Adige, or Tiroler Etschland. Before 1919, under Habsburg rule, the German-speaking population of this entire political complex on both sides of the Brenner Pass was called *Deutschtiroler*. After 1919, when Alto Adige and the Trentino passed into Italian hands, German speakers began to restrict the term "South Tyrol" to the predominantly German-speaking area in Italy between the Brenner and Salurn (Italian, Salorno). South Tyrol thus now explicitly *excludes* the Italian-speaking area of the Trentino.

The Trentino, before 1919, was also often known as *Welschtirol*, as contrasted with *Deutschtirol*. *Welsch*, often in use now by German speakers with a strongly derogatory connotation, traces its origins back to an ancient Indo-European root *walos*, that is, stranger, once applied by Teutonic tribes to both Celts and Romance speakers. (It survives today in such names as Welsh, Wales, and Cornwall in Western Europe, and as *Vlach* in the Balkans.) This term, *Welsch*, no longer has political currency, though it is still employed by German and Austrian historians writing about the formerly Italian-speaking component of the historic Tyrol. There are also some Italian-speaking inhabitants of the modern Trentino who, through continued fidelity to the Habsburgs, continue to refer to themselves as Trentine Tyrolese.

Locally—around St. Felix and Tret—there is some use of the purely administrative terms of "Provinz Bozen" (Provincia di Bolzano) to designate the German-speaking South Tyrol or Alto Adige, while "Trento" (Provincia di Trento, Provinz Trient) refers to the Trentino proper. This seems to be based on the villagers' differential involvement with the major administrative centers of the two provinces, Bozen/Bolzano and Trento/Trient, respectively. Before 1948, the German-speaking villages of St. Felix, Unser Frau im Wald, Proveis, and Laurein belonged administratively to the province of Trento; after that date they were transferred to Bozen.

We have tried to cope with the plethora of politically loaded terms by speaking of "the Tyrol" when we mean the undivided political unit of the pre-1919 past. When it is necessary to distinguish between the components of this unit to the north and south of the Brenner Pass, we speak of the cisalpine Tyrol as the area to the south of the pass, and of the transalpine Tyrol to the north of it. We use "South Tyrol" as isomorphic with the post-1919 Province of Bozen, and "Trentino" for the post-1919 Province of Trento. When we speak of "South Tyrolese," we mean the German-speaking people south of the Brenner; when we speak of "Trentini," we are referring to the Romance-speaking inhabitants of the Trentino. We give the German names of geographical features such as rivers, towns, or mountains when these occur within the confines of the Province of Bozen, and the Italian designations when these occur within the confines of the Trentino. As mentioned earlier, we have substituted, wherever possible, the name Anaunia for the German Nonsberg or the Italian Val di Non. We do, however, on occasion use the term *Nonsberger* for the German-speaking inhabitants of the Anaunia and Nónes for its Romance speakers. We have left untranslated a very occasional use of the German term *Deutschgegend* (German area) for the four German-speaking communities of St. Felix, Unser Frau, Laurein, and Proveis, which were administratively part of the Province of Trento before 1948.

QUESTIONS AND ANSWERS

As we look back upon our study, we see that we have been engaged in a dialectic between our somewhat divergent research interests and modes of interpretation. Wolf began the work with a concern for cultural differentials and sought answers primarily by asking questions about social and political organization; he was concerned only secondarily with ecological and economic problems. Much of his work was taken up with the eliciting of genealogical material and with the study of the church archives on baptisms, marriages, and deaths. Yet the inquiry into family histories inevitably drew him into study of inheritance and succession, and their implication for the continuity or dismemberment of homesteads in the two communities. By the end of the field work in 1961 Wolf had reached the conclusion that the cultural differences between St. Felix and Tret could be formulated primarily in terms of two major contrasts. The first, operative in the social microcosm of the two villages, could be discerned in the repeated assertion in St. Felix that its inhabitants favored impartible inheritance by the eldest son, whereas the Trettners were in favor of complete partibility. The second contrast characterized the relation of each village to the social macrocosm in which it was encased. The Trettners sought the largest possible involvement with the urbanizing and modernizing outside world, whereas the St. Felixers clung steadfastly to their identity as cultivators and equated that identity with their larger cultural continuity as Tyrolese or non-Italians. The two contrasts, microcosmic and macrocosmic, appeared to be related in that impartibility in inheritance appeared to favor continuity upon the land, while partibility seemed to favor mobilization, or dispersal, of resources within the context of the larger urban-oriented market. Wolf thus completed the first part of his field work with the hypothesis that the key to an understanding of the cultural differential between the communities lay in their different systems of inheritance.

It was the theme of inheritance that Cole selected for closer analysis and that formed, after his first fieldwork, the subject of his doctoral dissertation. He began, however, with a primary interest in ecology and economics, rather than in social organization and politics, and his questions were initially directed more toward identifying and analyzing the characteristics of ecological adjustments in the two villages. His questions and answers led him in the opposite direction from that taken by Wolf—toward an underlining of the similarity in ecological adaptation in the two villages and of the convergence of their actual practices in inheritance. We thus were forced by our data to acknowledge, as Edmund Leach (1961) had done before us, that social structure—conceived as a template of ideas for the ordering of social life—and actual practice, as apparent in the data when ordered numerically, were to a surprising degree antithetical.

We thus have been led from the initial hypothesis emphasizing differentials in inheritance systems to a more complex view that notes convergence between the two villages in ecologically grounded practice and divergence in politically grounded ideology. We note that different political orientations, productive of divergent ideologies, react in turn upon ecological processes, to produce, in the present, divergent reactions to expanding economic opportunities within both villages. In the course of our work we have come to see the relation between ecology and politics as a complex dialectic, rather than as a mechanical sum of significant variables. In this book we shall explore the interplay between thesis and antithesis. People live their local ecology and they also live in terms of social and political commitments to a larger social and cultural orbit. If we have not solved the problem of why people are able to combine such mutually inconsistent involvements, we are, at least, now better able to specify the nature of the problem.

We thus discovered, in the course of our study, a disjunction existing between the processes growing out of the local ecology and aspects of the local cultures that seemed to originate in their relations to the "outside" world. We therefore were led to try to specify the characteristics of that world. That endeavor, in turn, soon led us away from assumptions underlying much recent work by anthropologists interested in the study of complex societies.

The study of communities in complex societies was originally modeled on the study of isolated tribal groups. The temptation to treat these groups as "closed" systems was further reinforced by views that held that the community constituted a replica of the nation writ small, containing within its boundaries the significant features of the nation writ large—just as the homunculus in the male sperm was once thought to contain all the features of the future adult. Closely related to the temptation to think of the community as a closed and typical system was the tendency to elevate the explanatory sketch of *functional* relationships developed in the course of fieldwork into a *causal* model. In this causal model, the features selected for attention were conceptualized as the components and mechanisms of a small machine. The community, as characterized by the fieldworker's sketch of functional relationships, thus came to be conceived as a closed system driven by its characteristic small machine. The role of the anthropologist, in consequence, became defined as the study of such small, closed systems and of the little machines that drove them. The workings of the larger whole, the enveloping society, was relegated as subject matter to other social scientists.

Such an approach, however, begs the question of how small social and cultural systems are related to the larger systems of which they form a part; and it precludes further questions as to how these different phenomena determine each other. It is soon obvious in fieldwork that the community is neither a closed system nor a homeostatic machine: the functional relationships that obtain within it are relations of adaptiveness and congruence, and not the causal components of a homeostatic motor. Underlying any process toward adaptation or congruence are causal impulses that flow from the requirements of the physical environment, on the one hand, and from the forces at work in the larger world, on the other. In complex societies, these larger, "external" forces often dominate and reshape the forces at work in creating the local ecology.

What one sees in the local community, then, is the outcome of two sets of forces, ecological on the one hand, economic, political, and ideological on the other. The resulting interplay at the local level influences not only what goes on "on the ground"; it also influences the nature and capacity of the larger system in the "outside" world. The characteristics and capabilities of such a system depend directly upon the successful or unsuccessful outcomes of these local interplays. In this view, neither the local system nor the larger world can be understood as if each constituted a closed system, connected to the other only by some mechanical umbilical cord. They are products of the shifting relations obtaining between them, and hence cannot be understood without an attempt to understand these relations.

The investigator thus requires a discovery procedure that will allow him to identify these relations and to determine their relative strength. We found that the best way of grasping their characteristics and their capabilities was to see them in the process of development, from a point in time when they were weak or absent, to another point in time when they had grown strong enough to affect local outcomes. In other words, we needed to think in historical terms, to visualize the relations of St. Felix and Tret to the Anaunia, the Tyrol, Italy, Germany, and Europe as a whole in historical perspective.

The reader therefore will soon discover that we promised him a study of two communities, but that we have set that study within the framework of a more general history. Should he not want to be encumbered by this concern with the past, he may skip directly to Chapter VI. Yet we think that a certain kind of history is essential in our task of explicating the small, mountainous universe of the Upper Anaunia. We are not interested in history conceived as "one damned thing after another," but in a history of structures relevant to the Anaunia, in their unfolding over time, and in their mutual relationships.

There indeed have been attempts to short-circuit the need for historical understanding by introducing a falsely schematic sociology that counterposes "traditionalism" with "modernity," and assigns failing or passing grades to particular societies and cultures in terms of their location on one or another end of a scale of "modernization." We do not want to follow this approach, which seems to us to labor under the handicap of misplaced abstraction. The terms in which it formulates its problems and methods fail to account for the ways in which "modernization" and "tradition" interpenetrate and determine each other; "modernization" is often mere "tradition" in overalls, and "modernization" often dons peasant dress. Nor can one short-circuit the intellectual process of understanding complex societies by merely speaking of them as "larger systems" without taking note of their internal morphology, for they exhibit both spheres of patterned coherence and spheres of contradiction and disjunction; they contain autonomous, peripheral, and secessionist spheres, as well as those held in tight control. These spheres and their organization—their location, scale, and scope within the total society—have their own "structural" history of growth and development, of integration, and frequently of disintegration, too. Most important for the understanding of such societies is the manner and timing of their "crystallization": in Chapter II we make the claim that one cannot understand the Tyrolese of St. Felix without knowing something of the time and manner in which the Tyrol as a whole achieved its characteristic cultural profile, nor can one claim to understand Tret without knowing the means and ends of the Italian *Risorgimento* and its burning desire to reconquer the "lost" provinces in the mountains from its Austro-Hungarian enemies.

Nor do "larger systems" exist in isolation from each other: their very growth, crystallization, and demise depend only in part on their internal course of organization building; the "larger system" is also the product of political and ecological movement by which nations in the process of industrialization develop their economies, politics, and ideologies in response to the successes and failures of their predecessors. We certainly could not understand the forces at work in the encounter of Tyrolese and Trentini without some understanding of the interplay between systems located to the north and south of the Brenner.

Finally, while the combination of factors on the level of the national or international system depends upon the ways it can affect localities, it is not isomorphic with the combination of factors on the level of village or valley. What macrosystem and microsystem offer each other and demand from each other is of necessity different: this kind of transaction, too, has a historical dimension, as each sequence of pressure and response determines the sequence of pressure and response to follow. We have written history not for its own sake, then, but to explicate the Upper Anaunia, one valley in its structured interplay over time with forces emanating from the outside world.

If we had listened better at the outset of our work, we would have discerned these lessons in a story told by the octogenarian curate of St. Felix about some of his tribulations during World War II. Like most German Tyrolese village priests, he was of peasant stock and lived among his parishioners as a peasant among others. He was also strongly dedicated to the maintenance of Tyrolese identity in the South Tyrol and to the continuation of the bond between that Tyrolese identity and Catholicism. He thus opposed both the Nazis and the Italian fascists; at the same time, he was a saintly man, beloved by all in the region, Germans and Italians alike. During the war, like every other peasant, he killed his pigs and hung them in his chimney to make smoked bacon. But there was a war on, and all meat was handed to the occupying German authorities for distribution through the ration system. One of his parishioners, a fanatical Nazi, heard that the curate had slaughtered a pig and denounced him to the German command post located in Unser Frau. Thereupon the Germans came, located the pig in the curate's chimney, confiscated the smoked remains, and announced a date for public sale. On the day of the sale, the curate's faithful parishioners repaired to the command post in Unser Frau, bought their rationed shares of the pig, and returned the sum of pig shares to the priest. But the story did not end there. At the war's end, the partisans from Tret who knew the accuser's identity broke into his pigsty and carried off one of his own pigs to replace the pig taken from the curate. Up the road they marched toward the church, red bandannas around their necks and armed to the teeth, carrying the Nazi's pig between them. The curate, in turn, explained that he could not accept stolen property. The pig was left in his pigsty, and a message went out to the Nazi to retrieve his pig.

The story presents a number of familiar themes: the curate in St. Felix, as in most Tyrolese villages, is *primus inter pares*, a peasant like any other, entitled to smoke and consume his own pig. Outsiders, especially those in uniform, should not be allowed to interfere with their rules and regulations in what are essentially village affairs. For both Germans and Italians, the priest is a special personage; even partisans, in a situation of group conflict, could act in terms of the legitimacy of the priestly role, held in this case by a good man, though *todesk* (German). Thus, in this isolated valley the transcendental issues of war and fascism, national liberation, and of religious faith were transmuted finally into ecological counters—pigs confiscated and pigs returned. What we have learned in the course of this study is that the "imponderabilia of daily life" and the greater agonies of the human spirit are not divergent aspects of reality; they interact within one constellation that bears the hallmarks and stigmata of larger contradictions. It is with this realization that we ordered and thought through our field materials.

CHAPTER II

The Forging of Tyrolese Identity

The ancient land of the Tyrol is poised upon both sides of the Brenner Pass, which now connects Austria and Italy across a narrow saddle at 1370 meters above sea level. A unified political entity since the early thirteenth century, the Tyrol has been divided since the Treaty of St. Germain of 1919 into the transalpine Tyrol, which forms one of the provinces of the Federated Austrian Republic, and the cisalpine Tyrol, now organized as the region of Trento–Alto Adige of the Republic of Italy. Located at the interface between Mediterranean Europe south of the Alps and continental Europe to the north (see Figure 2), the land in the mountains has long been exposed to the advances and recessions of cultural tides from both north and south, advances toward the heads of the mountainous valleys, and renewed recessions toward the surrounding piedmonts and flatlands. At the same time, its mountainous perimeter has enabled the Tyrol to cling to ecological adaptations already made, while allowing the gradual assimilation of new cultural patterns. Changing

Figure 2. Places and passes in the Alps.

relations with the outside world, alongside steadfastness to its old ways, have characterized the Tyrol since ancient times.

THE ALPINE LANDSCAPE

The interplay between external and internal forces is strongly influenced by geological and geographic morphology. The central core of the Alps, made up predominantly of crystalline rock, runs straight through the center of the Tyrol along an east–west axis. At the center of this massif, the Brenner line forms a watershed between those waters that drain to the Inn River, and hence into the Danube and the Black Sea, and those flowing in a southerly direction, down the Etsch, or Adige River, into the Adriatic. The crystalline central core of the Alps is flanked to the north and south by ranges composed mostly of structurally weaker sedimentary rock. While the central massif gives rise to the highest Alpine peaks, in all areas of the Tyrol there are impressive variations in altitude between valley floor and mountain top. South of the Brenner, in the Trentino, the Adige descends to 200 meters above sea level, whereas the highest peak in the South Tyrol, the Ortler, climbs to 3899 meters.

The Pleistocene glaciers, which at times completely covered the Alps, have profoundly affected the potential of the mountains for human use. In the central core area, in particular, these frozen rivers of ice ground away many of the older soils, leaving exposed large areas of barren rock. As much as the resulting spectacular forest-free slopes attract the modern skier, they have proved of little value to permanently settled cultivators. The glaciers also widened longitudinal valleys, such as those of the Inn, the Eisack, and the Etsch, or Adige. Occasionally, a valley filled with water to form one of the much admired alpine lakes—Lake Garda on the Trentino's southern boundary is one of these. More commonly, however, they received a fill of alluvial deposits of both glacial and postglacial origin. Owing to their recent geological formation, the water courses that wend their ways through these deposits are rarely well defined; the soils in such valleys, therefore, are characteristically waterlogged. Thus, before the introduction of effective drainage techniques, many of the valleys contained bogs and swamps. Finally, the valleys are subject to annual flooding by waters drawn from the deposits of melting snow in the spring and by seasonally heavy rains in the fall.

The transverse valleys, in contrast, were not eroded by glaciers to anywhere the same extent. Consequently, they typically overhang the main longitudinal valleys to which they are tributary. Rivulets that have cut through these high valleys since the glacial epoch have tended to produce deep, narrow canyons or gorges through which they rage to join the larger rivers in the valleys below. Human movement through the gorges is, at best, difficult, and frequently travel is impeded by seasonal floods, landslides, avalanches, and similar disasters. Thus, natural barriers and disasters frequently cut off the transverse valleys from the adjacent lowlands. We shall have occasion in later chapters to further examine the ecological implications of these geological and geographic characteristics for human life.

PREHISTORY

There may well have been people in these hills during the Paleolithic Age, but their presence to date is attested to, not by any meaningful assemblages, but only by scattered artifacts. Yet, cultivators of domesticated grain crops and keepers of domesticated livestock were already well established in the Danubian lowlands to the east and in the alpine foreland to the west (see Higham 1967), when the first permanent colonists made their way into the high mountains. These forays were well under way by the year 2000 B.C., substantially aided by a climatic warming trend that steadily pushed back the forbidding ice of the glaciers, drove the tree line upward, and made possible settlement at higher altitudes.

The first attempts at creating fixed habitations in the high Alps were reinforced by attempts at large-scale copper mining, prompted no doubt by the growing use of bronze in southeastern Europe. In the Val Camonica, just to the southwest of the Tyrol proper, copper miners created a flamboyant culture. These people were quite possibly hunters and gatherers, who combined a gathering way of life with mining; or perhaps cultivators who farmed farther downhill and sent parties into the mountains to mine and hunt. On the Kelpachalpe, in the community of Aurach, near Kitzbühel in the transalpine Tyrol, the remains of mining activities are found in association with large quantities of broken bones belonging to domesticated animals (Pittioni 1954: 529–530). Such variable associations of metal work and food production also characterized the spread, somewhat later, around 1200 B.C., of the Urnfield culture, so-called because its bearers practiced cremation and burial in urns. More important than the burial practices of these people, however, was their accentuated mobility, based on horse-drawn vehicles, their military prowess, as shown by the use of the slashing sword, and their wide-spread commerce in metal objects and salt. The florescence of this culture demonstrated, to a high degree, the "independent potentialities of European peasants at a time when civilization in the Eastern Mediterranean was undergoing a temporary eclipse [Clark 1962: 155]."

The Urnfield radiation, in turn, produced numerous local and regional variations. Thus, Urnfielders fused with the bearers of an older Appenine culture to produce the Villanovan culture, in the area between the Po and Tiber rivers, while west of the Alps,

Urnfielders and preexisting populations produced a culture borne by Celtic speakers. In the alpine area proper, however, a basic Urnfield culture persisted in relative isolation. An alpine core, including the Tyrol, the Grisons in Switzerland, Liechtenstein, and Vorarlberg, remained the sanctuary of a culture that the Austrian archaeologist Pittioni has described as "the culture of Alpine continuity" (*Kultur der Alpinen Beständigkeit*). Taking up only occasional influences from the Etruscans and Hallstatt peoples, and little touched by the Celtic migrations, the hamlets and small villages of the alpine Urnfielders may well have lasted into historic times, with connections to the somewhat mysterious Rhaeti mentioned in Roman sources.

Little is known of the linguistic affiliation of these people, beyond the fact that they spoke some variant of Indo-European. Their linguistic heritage is evident today largely in place names such as Schlandraun, Taufers, Luimes, Nauders, and Tisens, to name only a few. Any hypothetical connection, once very popular, with the speakers of Illyrian (so-called after Illyricum, the lands immediately to the east of the Adriatic) should now be discounted. Nor is there more than a coincidence between the name Rhaeti and the Alpine languages now often grouped together as Rhaeto-Romansh. These languages, spoken today by some 450,000 people, are properly classified as Romance, regardless of whether one assigns them the status of separate Romance languages, or of variants of Italic. Geographically, they occur in three distinct areas: in the Swiss canton Grisons (Graubünden), where they are called Grisunsh or, officially, Romansh; in the Italian Dolomites east of Bozen/Bolzano, where they are called Ladinsh; and in Friuli, pivoted upon Udine. Undoubtedly, the area in which these languages were spoken was greater in the past than it is today; at one time they extended from the Grisons into the Inn River valley in the transalpine Tyrol, and into the Vintschgau/Val Venosta in the cisalpine Tyrol. They may represent an influx of Italic speakers from the south, or the effects of Italicization upon a preexisting population. All are marked by archaic retentions such as plural formations with -s; the use of diphthongs in the syllable-final position; and a high incidence of *pl*-, *kl*-, and *bl*-, as contrasted with the more frequent palatilizations of *pa*-, *ka*-, and *ga*- in Italian and Romanian.

Friuli was conquered by the Romans as early as 220 B.C.; the Ladinsh and Romansh areas, not until 15 B.C. Romanization, in a cultural sense, was probably confined to strategic places and transportation routes. Thus, at Salurn/Salerno in the cisalpine Tyrol, the

Roman road to the Brenner also traced the route of Romanization; but a few miles away the local population used a cemetery into the fourth century A.D., which exhibits completely pre-Roman patterns (Egger 1965: 25). The situation suggests the conditions of more recent times in which an Italian urban and bureaucratic culture is dominant in the strategic cities and transportation sites of the lowlands, whereas the mountains remain in the hands of a rural and tradition-oriented population.

NEW SETTLERS

The disintegration of the Roman Empire into a series of successor states unleashed a new advance into the mountains and across their passes, this time from the north. The western Rhaeto-Romansh-speaking area was invaded by Alemanni around the end of the fifth century A.D. Friuli lay across the invasion route of the Visigoths and Ostrogoths, and later fell to the Longobards. The Tyrol was occupied largely by the Bavarians. Taking advantage of the weakened grip exercised by the southern lowlanders upon the mountains, and spurred on by Frankish kings who hoped to enlarge their perimeter of defense, Bavarian nobles moved southward during the seventh and eighth centuries; by A.D. 680 Bozen/Bolzano had a Bavarian count.

At first slowly, then after A.D. 1000 in ever widening circles, the new settlers and the descendants of old settlers under Bavarian aegis expanded from the original terrace settlements above the valley floors into the higher altitudes. Nobles, settled in their newly constructed castles, and ecclesiastical organizations, were the primary sponsors and catalysts of the movement. Thus, the bishopric at Brixen/Bressanone was in all likelihood responsible for the creation of the present Ladinsh-speaking area in the eastern Dolomites, when it prompted the movement of Romance-speaking settlers from the valley of the Eisack River into the mountainous uplands (Schurr 1965). Elsewhere, mountain land was taken over by German speakers, in an expansion of populations of German speech that extended far beyond the boundaries of the cisalpine Tyrol proper. In what is today Switzerland, Alemannic speakers pushed southward in the eleventh or twelfth centuries into the otherwise French-speaking Valais, into the area around the Monte Rosa, and into the present-day Italian regions of Val d'Aosta and Piedmont. In the now largely

Italian-speaking Trentino, German settlers were welcomed as miners and cultivators into the valleys of the Avisio River and into the Val Sugana (see Figure 3). German-speaking linguistic islands (in Brandtal/Vallarsa, Laimtal/Terragnola, Vielgereut/Folgaria, and Lusern/Luserna) trace their origins to the thirteenth century. During the same period, German speakers were settled in the now Italianized Seven Communes of Vicenza, with their center at Slege/Asiago, and in the Thirteen Communes of Verona (*Dreizehn Kamaun von Bern*) (see Table 1).

Since many river valleys in the initial area of colonization were subject to flooding and were badly drained, relative overpopulation on the narrow valley terraces provided a continuous demographic stimulus to expansion (Wopfner 1951–1960, Vol. I: 65). A drive into higher altitudes was already under way in the transalpine Tyrol in the eleventh century, and in the Tyrol south of the Brenner in the twelfth. At first many new homesteads were livestock holdings (*curtes stabulares*), frequently erected under noble or clerical sponsorship. Since the twelfth century such holdings are called *Schwaigen, Schwaighoefe*, after the Middle High German *Schwaige*, signifying both herd and pastoral holding. Overlords supplied the initial stock of animals and at first furnished the needed grain. Later, many of the formerly purely pastoral holdings began to grow their own crops, and by the fourteenth century many of them paid the tithe in wheat or oats.

In contrast to earlier settlements in densely concentrated villages, the colonized uplands were usually settled in isolated homesteads. This pattern of settlement is sometimes regarded as typically Germanic, in contrast to a putative Romance preference for densely clustered settlement in villages, sometimes attributed to a supposed pattern of Mediterranean sociability. Isolated homesteads, however, correlate more with colonization than with ethnicity. They predominate among the Romance-speaking Ladinsh of the eastern Dolomites (Wopfner 1951–1960, Vol. I: 89; Lutz 1968: 104), and occur in Germany proper, in "new" areas of settlement, rather than in the "old" habitation zones where concentrated settlement prevails (Wolf 1970).

The colonization movement affected, in important ways, the dues and rents collected by the overlord. Everywhere the presence of a land frontier appears to have been instrumental in easing rural rental agreements (see Chapter IV). Increasingly common—and soon to become widespread in the German-speaking zones—was the grant of hereditary tenure against the promise of hereditary rental

Table 1

	Europe	
Early Middle Ages A.D. 600–1100		
Late Middle Ages A.D. 1100–1500		
Early Modern Period A.D. 1500–1700	1520	Luther launches the German Reformation
	1525	German peasant war
	1543– 1563	Council of Trent
Modern Period A.D. 1700–Present	1718	Trieste becomes a free port
	1796– 1814	French invade Italy
	1861	Kingdom of Italy created
	1915	Italy enters World War I
	1919	Treaty of St. Germain
	1922	Fascist seizure of power
	1938	Hitler occupies Austria

(*Erbzinsrecht, locatio perpetua*). Rents were fixed; anything above the agreed limit was the property of the pastoralist or cultivator. Heirs, through lineal descent, inherited the agreement; collaterals were excluded, as were ascending lineals, from inheriting. The peasant could sell the holding with the consent of his lord, who had the right to make the first bid for purchase; however, the holding could not be divided without the lord's express consent. The lord undertook to stock the holding with animals and supplies; in case of poor maintenance, he could sue for its return. Legally, the arrange-

CHRONOLOGY

	Tyrol		Anaunia
680	Bozen has a Bavarian count		
1040	Counts of Tyrol become lords of the Vintschgau		
		1184	First documentary mention of Unser Frau
1254	Tyrol unified under the Counts of Tyrol		
		1284	First documentary mention of Tret
		1342	First documentary mention of St. Felix
1363	Habsburgs fall heir to the Tyrol		
		1407	Peasant rebellion in the Anaunia
		1477	Peasant rebellion in the Anaunia
1525–1526	Tyrolese peasant war	1525	Peasant rising in the Anaunia
1809	Tyrolese uprising against French and Bavarians	1809	Uprising in the Anaunia
1919	South Tyrol and Trentino become Italian		
1939	*Option* organized		
1945	South Tyrol again Italian		

ment left the peasant "free," in the sense of freedom from the lord's political and legal jurisdiction. From very early times on, therefore, the colonization movement and the beneficial tenure arrangements stimulated the growth of a politically free peasantry. In this, the Tyrol was ahead of neighboring areas; the independence of the peasant was to remain one of the salient characteristics of the Tyrolese population up to the present day. A parallel movement from serfdom to free contract also took place in northern Italy and in the Romance-speaking areas of the Tyrol. Yet here the dominant

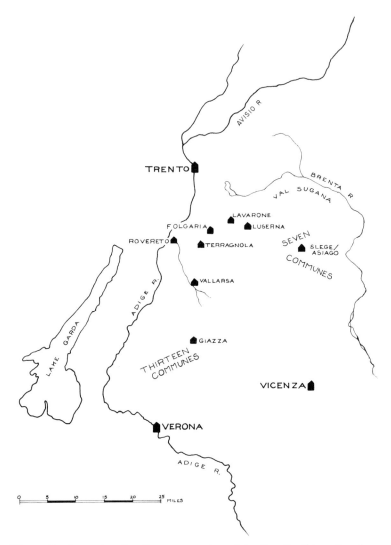

Figure 3. Expansion by German speakers into the Southern Trentino.

form of tenure remained temporary rather than hereditary (Wopfner 1951–1960, Vol. III: 469).

Agricultural expansion was accompanied by commercial expansion. With the Tyrol encompassing the two passes of Brenner/ Brennero and Reschen/Resia, the Roman roads over these passes acquired renewed importance in linking cisalpine and transalpine Europe after the political and economic consolidation consequent upon the growth of the continentally based Carolingian Empire. Political motives for the renewed use of these transit routes may well have predominated during the early Middle Ages, when the Holy Roman Emperors repeatedly attempted to convert the ideal construct of a Roman Empire of the Germans into political fact through repeated invasions of Italy. The first to use the Brenner thus was Otto I in A.D. 951; the largest army to cross the pass, led against Milan by Frederick Barbarossa in the twelfth century, consisted of 100,000 foot and 15,000 horse (Pivec 1961: 94; Widmoser 1961: 304). Yet, until the thirteenth century, the Brenner was one pass among many, and the Swiss passes of the Septimer and the St. Bernard serviced the economically strategic traffic between the "vital hinge" of the Champagne fairs and Italy. In the thirteenth century, the growing prosperity of the Italian towns reversed the field of force, and the Brenner came into its own as the vital channel between southern Germany and the Italian plain, and—most especially—with Venice. Venice exported cottons and silks, spices, and glass; in return, it imported, from the north, linen, wool and furs, pigments, iron tools, and precious metals (Simonsfeld 1887, Vol. II: 103, 197).

This growing commercial intercourse had a special effect on the Tyrol, through which the major share of the associated traffic passed. Directly, it stimulated the rise of Tyrolese towns, most of them founded in the thirteenth and fourteenth centuries. From the beginning they were oriented toward trade and transportation; very few of them had any connection with earlier Roman or monastic settlements (Metz 1961: 343–344). Taverns and hostelries grew in large numbers and remain a striking feature of the Tyrol. Their characteristic artisans were smiths, carters, and wainwrights, rather than weavers, who were important in the towns of southern Germany and Italy; weaving in fact remained a purely rural pursuit in the Tyrol. The growing towns, in turn, needed supplies of food from the surrounding countryside, thus widening the internal market and promoting the use of money in commercial transactions. Competition from the towns for labor and resources undoubtedly reinforced the pressure for easing rent agreements in the countryside, and it intensified still further the shift toward a free, food-producing peasantry.

POLITICAL CONSOLIDATION

A second consequence of the growing transit trade—as much cause as consequence—lay in the political consolidation of the Tyrol. Eleven different political units were fused into one under the

Counts of Tyrol, who had their seat in the somewhat peripheral Vintschgau/Val Venosta athwart the road from the Reschen to the lower valleys of the River Etsch/Adige. From the thirteenth century on, the Counts of Tyrol worked diligently to curtail the political and juridical power of the nobles, both through the appointment of nonaristocratic judges and administrators, and through the increased participation of the peasant estate in political decision-making. After A.D. 1415, the peasantry was represented in the Tyrolese assembly, a rare event in late Medieval Europe. Moreover, the Counts pursued a policy of furthering peasant economic independence from overlords by granting favorable rights of hereditary tenure in land. This had two important consequences. First, the Tyrolese nobility never developed any overreaching powers of its own; its sphere of influence remained largely local. Second, whereas

Castle Tyrol near Meran, where "the Counts of Tyrol had their seat athwart the road from the Reschen to the lower valleys of the River Etsch."

the Counts of Tyrol favored the peasants against the nobility, the peasantry achieved its freedom and its rights through collaboration with the central ruler, and not—as in Switzerland—through the independent association of its communes (Mayer 1959: 239–240). The concept of Tyrolese freedom thus came to include the legitimacy of the dynasty, a political benefit inherited by the Habsburgs when they succeeded the Counts of Tyrol in A.D. 1363. Like their Tyrolese predecessors, the Habsburgs maintained the balance of social forces through a reliance on a free peasantry, and they proved able to accept a framework of representative government in the Tyrol, which they were quick to deny populations in other portions of their growing empire.

Habsburg ascendancy further emphasized the role of the Tyrol as a connecting link between adjacent areas, this time not only between north and south, but also between their western outreaches and eastern possessions (Sölch 1923). It must be remembered that the Habsburgian transalpine Tyrol was connected with the rest of Habsburg Austria by the Brenner Pass and the valley of the Puster River/Pusteria. This valley, of vital importance in Habsburg empire-building, was acquired by Maximilian in A.D. 1500 and united with the Tyrol:

> Lord of the land of Tyrol was not whoever had control of Brenner as such, and thus controlled traffic from north to south, from northwest to southeast, from east to west, but he who held in his hands all the access routes, from the points of entry for traffic from the surrounding flatlands into the mountains, he who was in a position to close off the roads leading through the valleys of Lech, Isar, and Inn in the north, the gorges on the Etsch, Brenta and in the vicinity of the Toblach plain. Only from these requisites could one explain the shape of the Austrian Tyrol [Sölch 1923: 26].

TRADE AND MINING

At a certain point, however, increased commercialization due to long-distance trade and the rise of towns began to have a negative effect on the ecology of mountain farming. It has been suggested by some historians that climatic conditions became less favorable after A.D. 1300, an important factor in marginal mountain agriculture (see Le Roy Ladurie 1971). Population rose, especially in the areas of "older" settlement in river valleys and on valley terraces; there was also a marked tendency to group single homesteads into hamlets and hamlets into villages. Wopfner (1951–1960, Vol. II: 223) has suggested that the population rose by one-half between A.D. 1312 and 1427. We estimate that it rose from roughly 240,000 in 1312 to about 360,000 in 1427. Pastoral holdings were increasingly converted to agriculture in order to provide food (Wopfner 1951–1960, Vol. II: 136). Division of holdings became more common, especially in areas of traditionally partible inheritance, like those of western Tyrol and the Trentino. With increased population, overlords may have encountered difficulties in collecting the dues owed them; at the same time, generally depressed prices within an increasingly commercial agriculture, a wider European phenomenon, may have prompted them to increase pressure on the peasantry. Political centralization exacted its own costs in increased taxes and levies in the Tyrol, as elsewhere.

To the variable effects of demographic pressure, commercialization of agriculture, long-distance trade, and the growth of cities were added, in the fifteenth and sixteenth centuries, the effects of a Tyrolese mining boom. Some mining activity had gone on since prehistoric times, and a few mines were to continue producing into the modern period. But in the fifteenth century the production of silver and copper took a sudden leap, ending with the virtual exhaustion of deposits in the early 1600s. For this brief period, the Tyrol provided a substantial part of the wealth that filled the Habsburg coffers and underwrote the pan-European ascendancy of the dynasty.

In the Tyrol itself, mining had far-reaching effects, creating a kind of "premature" capitalist development. Much of the capital for the new mines was furnished by capitalists from southern Germany, like the Fuggers; mining came virtually to a standstill when the spreading bankruptcy of the Spanish and French states also destroyed the fortunes of the Fugger bank. For the first time, the Tyrolese witnessed the growing power of a class of foreign entrepreneurs. On the other hand, mining created a sizable wage-earning proletariat, many of whom were recent migrants from Germany, and brought new skills, but also competed with local miners for available jobs. According to Correll (1927):

> Rapid economic progress was responsible for the social pressure which early weighed heavily on Tyrol. The spread of the money-economy and the beginning of the capitalistic social order resulted in foreign financial control of the ore mines. The native miners were gradually excluded from their ancient guild mining privileges and suffered from the competition of foreign miners who migrated as common laborers into the Tyrol mining

regions. Vested feudal interests and the profit urge of international investors competed for returns [p. 49].

Miners, in turn, brought dependents. In Schwaz, the main mining center of transalpine Tyrol, there were said to be 10,000 miners; their dependents have been estimated at 30,000 (Tremel 1969: 178). Similar numbers are reported for Sterzing/Vipiteno and Gossensass/ Colle Isarco, the two mining centers of the South Tyrol (Geizkofler 1961: 132). Mining, in turn, increased economic demand with its dependence on transportation, lodging, commerce, food supplies, leather, wood, and charcoal. Moreover, many miners acquired agricultural holdings, buying and selling land for subsidiary agricultural pursuits, and thus accelerated both the division of holdings and population pressure upon the land. In the early sixteenth century, both the pressures upon the peasantry and the social realignments due to mining began to generate movements in opposition. Some of these were openly rebellious and issued in the Tyrolese Peasant War of 1525–26. Others were ideological, and preached the advent of a new religious order.

THE PEASANT REVOLT

As elsewhere in the German-speaking areas of Germany, Switzerland, and Austria, the peasant uprisings that marked the second quarter of the sixteenth century were essentially conservative— assertions of inherited rights against the encroachment of overlords and against "petty civil and ecclesiastical lords . . . attempting to extend and formalize their own jurisdiction at the expense of the peasants [Williams 1962: 61]." In the Tyrol, this was reinforced not only by a growing hatred of foreigners such as the Fuggers, but also by the accession to power within the Habsburg realm of the Spanish line of the dynasty, and most especially by the employment, in the Tyrol, of Spanish bureaucrats like Gabriel Salamanca, whom the Tyrolese characterized "smelling, heretical, asarianic [sic!] Jew and villain."

It is probably no accident that the revolt had its origin in the bishopric of Brixen/Bressanone, where the ecclesiastical establishment strove to introduce a new antitraditional rationalism into its various affairs. In this region, moreover, a semiserf population working in the vineyards of the Bishop played a role in giving the revolt a more radical political cast (Maček 1960; Franz 1965: 292). A radical assembly in Meran demanded unification of the legal system; freedom to preach the word of God; secularization of church properties; the dissolution of monasteries and nunneries; control of the tithe by laymen and its use for the upkeep of the poor; protection against high prices for shoddy goods; the subjection of nobles to the codes of the communities where they lived; the right of communities to elect their own judges and priests; the restriction of peasant dues to rent and tithe, and freedom from work dues; an end to serfdom where it still existed; free access by all to pasture and forest; an end to guilds; an end to exploitation by foreign associations, such as the Fugger bank; an end to jailing without legal procedures.

The leader of the rebellion, Michael Gaismair, even envisaged the founding of a peasant republic without king, nobles, or clergy, to be administered by elected judges—a state without merchants and usury, in which mines and workshops were to be nationalized and foreign goods barred by high tariffs. The revolt, which began at Brixen, quickly spread to Bozen and Meran, then into Anaunia and the Val di Sole, the Val Lagarina, and the Val Sugana. The radical proclamations were translated into Italian and read to the Italian-speaking population. But the rebels were unable to take Trento. The population of the towns grew frightened by peasant demands for the abolition of guilds and usury; the transalpine Tyrol remained quiescent. Neighboring Bavaria remained a "redoubt of quiet and thus of reaction [Franz 1965: 289]." Agitation among the miners was met by concessions, a move which succeeded in separating them from the peasant rebels. At the Tyrolese assembly in 1526, the duke accepted many peasant demands, only to rescind them again when peace was restored in 1532. Michael Gaismair, after attempting a renewed invasion of the Tyrol from Switzerland, fled to Venice, where he was murdered by an imperial hireling in 1527. His utopian plan for a peasant republic based upon God's word, in which the "cities were to be reduced to villages [Franz 1965: 158]," was drowned in blood.

A parallel fate befell the ideological rebellion, which in the Tyrol, more than in other parts of Austria, took the road of what George H. Williams has called "the radical reformation." Especially significant in this movement were the Anabaptists, a congery of religious sects holding common beliefs in a return to a pristine Christianity with a commonality of goods, but without either state or clergy, and who symbolized in their double baptism their "voluntary separation from the unredeemed world [Cohn 1961: 274]." While most Anabaptists were fiercely pacifist, it is probably more than a symbolic coincidence that Michael Gaismair's brother, Hans, was executed as a baptist in 1526. The political rebellion and the ideological insurgence appear as phases of a single, overall movement. In Williams' words (1962):

> The importance of foreign miners in the newly worked Fugger holdings in these valleys, the consequent breakup of customary social securities, and the distention and consequent understaffing of the old medieval parishes inundated with immigrants constitute the social and economic matrix of Austrian, and particularly Tyrolese, Anabaptism, and before that of the peasant uprisings in the same areas. Anabaptism was, in fact, 'the core of the Protestant revolution in Tyrol.' Revolt and Anabaptist evangelicalism seem to have been successive solutions or responses to the same social disturbance in the congested valleys, but once converted to Anabaptism, the Austrians were nonresistant to the point of being almost eager for martyrdom. Austrian Anabaptists appear to have been recruited from among all classes. Artisans and miners were much more numerous in relation to peasants than in other sections of the Empire [p. 166; see also Correll 1927 and Dedic 1939].*

In general, one has the impression that the political movement was more purely rooted in the peasantry, while the religious movement found greater acceptance in the incipient proletariat of the mines, among craftsmen, and among the "free-floating intelligentsia" formed from a variety of social groups attracted to the movement. Perhaps, also, the defeat of the political insurrection created the conditions for a popular millenarianism as an ideological substitute. Whatever the explanation, conversions to Anabaptism were numerous. The great extent of this conversion

> . . . has been obscured for a long time because of the very great success of Archduke Ferdinand and later the Counter Reformation in scouring clean of all religious heterodoxy the mining valleys that once teemed with Anabaptists [Williams 1962: 166].

While most Anabaptists, including those of the Tyrol, were pacifists, this was not true of all. The attempt of an extremist group to create in 1532–35 at Münster a communist, polygynous, Maccabean theocracy reinforced the perception of the secular power that the Anabaptists constituted a direct threat to constituted authority in state and church alike. Leaders were executed, and large numbers of Anabaptists were expelled. Not a trace of them remains in the Tyrol today. The last descendants of the Tyrolese religious rebels

---

* From *The Radical Reformation* by George H. Williams. Copyright © MCMLXII, W. L. Jenkins. Used by permission of The Westminster Press.

are the followers of Jakob Huter who have become the modern Hutterites.

THE COUNTER-REFORMATION

The Tyrolese persecution of the Anabaptists represented but a first step in the Austrian Counter-Reformation. In the Tyrol itself the Protestant movement had been in the main a movement of the lower estates; the nobility remained predominantly Catholic. But in the other Crown lands of the Habsburgs, the nobility turned largely Lutheran, using its new faith "in order to sustain its local particularism by a freer and more rational religion against the papal and imperial universalism [Jászi 1961: 46]." There, not only were the nobles vastly more powerful than in the Tyrol, but they threatened to draw after them the burghers of the towns and the masses of the unfree, dependent peasantry. Against this threat, the Habsburgs mobilized all their wealth and power, and they proved successful in rolling back the Protestant threat largely because the nobles were too intent on maintenance of their own privileges to cement the interclass alliance which alone could have spelled success for their cause. They were unwilling, in the words of Robert A. Kann (1967: 259) to pay "the price in social concessions which the extension of religious liberties to markets and towns, and in the end to the peasantry, would have entailed." Its victory allowed the Catholic dynasty to uproot the Austrian and Bohemian nobility entirely and to create a new court nobility recruited largely from foreigners and military supporters, which in turn became the devoted instrument of dynastic centralism and absolutism. Thus, any hope that the Protestant rebellion could install representative institutions based on an alliance between the various estates proved to be in vain. Instead, a bureaucratic state, ruling over weak classes, was firmly established.

From this point on developed the important cultural opposition between the cosmopolitan court at Vienna and the life of the Austrian hinterlands. The court created a flourishing, international culture pattern,

> . . . built up of various elements of Spanish, Italian, French and Dutch life, but with Spanish ideas at the core of it; the whole knitted together by the bonds of the Catholic Church. This international, upper class civilization, essentially Latin, Mediterranean and Catholic, aristocratic, bureau-

cratic, clerical, brilliant but somewhat rootless, has been continually at
strife throughout Austrian history with the much more modest, rather
backward and inarticulate but deeply rooted German elements of life of
country and town which date from the era preceding the imperial great-
ness of the Austrian Empire [Borkenau 1938: 36].

In the orchestration of this pattern, the Church of the Counter-
Reformation played a vital part in asserting the universalist, cen-
tralist, and supranational interests of the Habsburgs. Much of the
reconversion to Catholicism was accomplished by outright force;
some of it was due to the ideological fervor of what Friedrich Heer
(1966) has described as a "Platonic totalitarianism," within which

> . . . all real colors, tones, themes, men, races and continents were to
> unite in the great world theatre of the total, united world Church. Play,
> work, prayer, celebration, erudition and charity were all to become part
> of religious service . . .

> All things and creatures, all good super-terrestrial and sub-terrestrial forces,
> all the living, the saints and the dead were to celebrate earthly life as an
> act of thanksgiving and a feast of the majesty of God. All this was set
> before the eyes of the masses by the baroque Rome of the Counter-
> Reformation. Lutherans and Calvinist *heard* salvation, religious noncon-
> formists *thought* about it, the Averroists *knew* it to be immanent in nature,
> the pantheists *felt* it in the universe, the mystics *tasted* it in all things.
> But the Roman-baroque Catholic heard, thought, knew, felt, tasted and
> *saw*. Above all, he acted it out in the world theatre of the Roman baroque
> which in a remarkable way sought to unite the concerns of all enemies
> [p. 291].

Such orchestration on the symbolic level was the ideological
counterpart of synchronization on the political level. Where the
Protestant Reformation had permitted a reduction in the power of
the luxury-oriented Renaissance state, and hence the replacement
of a parasitic bureaucracy in favor of a working bureaucracy related
to real economic needs (Trevor-Roper 1967: 89), the Counter-
Reformation permitted Catholicism to survive in its old home, at
the cost of stepped up investment in the old bureaucratic apparatus.
Says Trevor-Roper (1967):

> The Counter-Reformation, which may be seen as a great spiritual revival;
> a new movement of mysticism, evangelism, charity. But sociologically it
> represented an enormous strengthening in the 'bureaucratic' structure
> of society [p. 34].

In the Tyrol, the defeat of the peasant rebellion and the destruction of Anabaptism restored law and order. At the same time, the

Church carried the word of the new consolidation into the most remote hinterland. Nunneries and monasteries multiplied; seminaries were created for the training of priests; a university was founded at Innsbruck in 1669 and opened to the children of the peasantry, as well as to the sons of other estates. The enlarged ecclesiastical establishment served to increase social mobility, giving rise to the hope ascribed to every Tyrolese family to number at least one priest among its members. When local schools were created in the eighteenth century, they were to be staffed by graduates of the new clerical educational institutes. Church attendance was made mandatory, on pain of suspicion of Protestant heresy; religious brotherhoods were organized around devotions to the rosary or to scapulars; separate, special devotions were set up for married men, married women, boys, and girls. Popular missions, especially under Jesuit auspices, spread the religious message to the villages through visits, sermons, and religious exercises, and they popularized the Marian devotions and daily recitation of the rosary within the confines of the peasant household.

Churches and chapels were built, and ecclesiastical construction gave rise to numerous religiously oriented crafts, especially wood carving and painting. Passion plays emphasized the story of salvation. In this fashion, the great "world theater" of sophisticated Baroque Christianity forged a link to the Christianity of the folk. In 1796 the new religious unity was given expression in the official recognition of a traditional Tyrolese devotion to the Sacred Heart of Jesus, a cult that symbolically underlined the role of the Tyrolese as a chosen people through the postulate of a special "covenant" between the Tyrolese and Christ (Dörrer 1947). This concept has remained central to Tyrolese identity, and the lighting of so-called Heart of Jesus fires on all the mountain tops on the third Friday after Pentecost remains to this day a symbol of Tyrolese ethnic identity and political unity.

Whereas details of the reintroduction or reassertion of Catholicism in the villages are all too scant, it is more than likely that the new religiosity was of major importance in permitting the social reintegration of peasant communities after the dislocations of revolt and religious schism. It is also significant that by the end of the sixteenth century the decline of mining brought about a withdrawal of south German capital and a dissolution of the mining proletariat, thus removing some of the root causes of social agitation. Further relegation of the Tyrol to a marginal economic role within the Habsburg Empire also intensified the return to more local and parochial concerns. Thus, paradoxically, a retreat into a more rural life coincided with the Counter-Reformation, to produce, under Habsburg auspices, an approximation to the peasant republic "without towns" for which Michael Gaismair had given his life. As a result, it would, from that time on, represent a stronghold of conservatism and rural traditionalism, its face set against urban, innovative tendencies, whether they emanated from Vienna or from the Italian lowlands to the south. The Counter-Reformation won more than a battle: it marked the transition from the "ducal County of Tyrol" to "the Holy Land Tyrol."

ECONOMIC AND POLITICAL DECLINE

With the decline in mining, the Tyrol became an economic backwater. In agriculture, similarly, it became ever more dependent on the importation of grain from other areas, chiefly Bavaria and Hungary, despite the extension of agricultural production, especially after the regulation of Inn and Etsch and the draining of valley bottom swamps. This agricultural base, together with the imports, had to feed an ever expanding population, estimated at 500,000 in 1650 (Wopfner 1951–1960, Vol. II: 275); a figure that was to reach 593,000 in 1754; 719,000 in 1835; and 947,000 by 1910 (Fiebiger 1959: 15). Craft production, in the form of cottage industries such as weaving, hat-making, basket-weaving, embroidery on leather, the production of artificial flowers, wood carving, distilling, and others, took up some of the slack, but did not produce larger industrial establishments. By 1780, a census that did not include the bishoprics of Brixen and Trento counted 46,000 peasants with land sufficient for their needs, and 61,000 who had to supplement their income from agriculture with other activities (Wopfner 1951–1960, Vol. II: 359). The depopulation caused by ravages of the Thirty Years War in Germany caused many Tyrolese to emigrate permanently in search of new opportunities abroad.

Equally significant was the seasonal migration of agricultural laborers into surrounding areas; their number is estimated to have reached 30,000 a year in the eighteenth century (Wopfner 1951–1960, Vol. II: 396). Migration, however, did not alter the newly strengthened social structure of the rural areas. Rather, it acted as a safety valve, merely reducing pressures against it. It reduced temporarily the number of mouths to be fed in the Tyrol itself; cash income earned abroad, in turn, paid for the importation of grain. Traveling peddlers from the Tyrol also became a standard feature of the European scene. Nevertheless, a free peasantry, settled upon its own land, remained the structural basis of Tyrolese society. It maintained most of its impartible holdings into the twentieth century.

Large-scale, long-distance trade through the Tyrol from north to south continued, supported by legislation that exempted the Tyrol from the tariff regulations that increasingly bound the remainder of the Habsburg monarchy. Most of this trade, however, was destined for transit, and much of it went increasingly eastward to Trieste, as the axis of the empire moved ever further to the east. Whereas the line from Trieste through Carinthia, Pusteria, and Brenner to south Germany retained some importance, the more direct connection between Vienna and the Adriatic port began to overshadow the Tyrolese routes and brought about the decline of the once prosperous Bozen/Bolzano fairs.

Within the larger structure of the Habsburg Empire, too, the Tyrol became increasingly marginal and politically irrelevant. The Tyrolese still had a part to play in the Austrian wars of succession, which involved the dynasty with its competitors to the west, but the center of gravity of the monarchy was moving ever farther to the east, into the areas wrested from the Ottoman Turks. The Tyrolese militia might inflict defeat on the invading Bavarians in 1703, but the significance of a battle of Pontlatz was vastly overshadowed by the victory of the imperial armies over the Turks at Senta in 1697.

Not only did the country suffer economically under the effects of this shift, but its special historical and social characteristics came into increasing opposition with the general dynamic of the monarchy. By this time, the Tyrolese peasant had developed a tradition of freedom; he needed neither liberation from serfdom, nor an end to corvée labor, which oppressed the peasantry to the east. He had the right to bear arms and to serve in local regiments; the peasants in other parts of the monarchy were hapless conscripts. He jealously maintained his right to inalienable, indivisible landholding; the rest of the monarchy was characterized by latifundia or by peasant holdings subject to dispersal through partible inheritance. The Tyrolese regarded the monarch as successor to the Counts of Tyrol and the special guarantor of inherited freedoms; to the remainder of the monarchy, the emperor was the absolutist head of a vast, impersonal bureaucracy. The Tyrolese addressed the emperor with the familiar "Du"; to the others he was approachable only through the pomp and fanfare of the courtly ritual. The Tyrolese nobility was old in the land, largely rural, disinterested in court affairs, and lived among the peasantry very much as *primi inter pares*. The nobility of the Habsburgs was relatively new, lived at court, saw in Vienna its spiritual center, and drew its sustenance largely from huge, landed properties worked by serfs and—later, after 1848 —by masses of landless, migratory, agricultural proletarians. The Tyrolese noble, like the Tyrolese peasant, served in local regiments and militia; the court nobility preferred service in nonlocal regiments, especially in the Galician and Hungarian cavalry. When the pace of industrialization increased in the latter half of the nineteenth century, it affected largely Bohemia, the Vienna basin, Bratislava, and Salgotarjan in Hungary; apart from a few cementworks, the Tyrol remained largely rural, pastoral, and agricultural. In the Danubian provinces, industrialization produced a large proletariat with a socialist ideology; the Tyrolese remained staunch rural conservatives and Catholic.

The exceptional role of the Tyrol within this larger political and social conglomerate found its most accentuated expression in the Tyrolese fight against the armies of the French Revolution and of its heir, Napoleon. The Tyrolese militia fought off a French incursion in 1796–97, only to see the Tyrol abandoned to Bavaria in 1805. The Bavarian government, under French inspiration, abolished the Tyrolese constitution, appointed new and more numerous judges, introduced conscription, and imposed new taxes. Bavarian occupation also cut off the trans-Tyrol long-distance trade and impeded the seasonal migration of Tyrolese to neighboring areas. But the specific causes of the Tyrolese insurrection of 1809 can be found in the anticlerical legislation of the new government, inspired by concepts of French revolutionary rationalism and anticlericalism. Processions and rogations were prohibited, churches were secularized and their contents sold, and priests were hauled into the courts. Rumors spread to the effect that the Bavarians were going to put an end to celibacy, to institute divorce, and to abolish auricular confession (Hirn 1909, Vol. I: 103–105, 107, 123). In April, the Tyrolese rose in rebellion, led by the innkeeper Andreas Hofer. As in Catholic Spain, where the French invasion of 1808 unleashed a guerrilla war in favor of the decadent Ferdinand VII, who represented the true faith and its legitimacy against the *afrancesados,* so the Tyrolese rose against the godless French and their Bavarian allies. In three successive major battles the Tyrolese proved victorious, but they lost the

"In April, the Tyrolese rose in rebellion, led by the innkeeper Andreas Hofer." (Original in the Andreas Hofer Memorial Chapel, Sandhof, Passeier.)

fourth. Andreas Hofer was captured and executed on February 20, 1810, to become, in song and in effigy, the symbolic incarnation of Tyrolese liberty, religiosity, and loyalty to an emperor who did little to support the rebellion of his subjects.

The Tyrol thus moved into the Age of Nationalism with all the burdens that were the results of its paradoxical history. The crucial paradox of this history lies in the fact that it could not benefit by any "privileges of backwardness," but had to pay instead the price of a premature advance into modernity. It had been caught up in one of the early movements toward state-building in Europe; its peasantry had early achieved political and economic liberties not granted to many other European areas. Through participation in long-distance commerce and in the mining boom, the Tyrol had stood at the cutting edge of European capitalism; its peasant revolt had been both product and response to this forward thrust. Yet, when its localized revolt had failed in its attempt to create an autonomous, egalitarian, peasant republic in the Alps, it was transformed into a conservative and encapsulated component of a cumbersome and hierarchically organized polity that bent Tyrolese aspirations to its own economic, military, and religious ends. The peasant republic envisaged by Gaismair became the Holy Land

Tyrol defended by Andreas Hofer in the name of a Holy Roman Emperor and the Mother Church. As a result, new threats to the continued existence of the multinational and pluralistic Habsburg Empire would also, in the centuries to follow, subject the Tyrol to those threats and the attendant stresses of dissolution.

CHAPTER **III**

Torments of Nationalism

With Tyrol thus relegated to a position of economic marginality and political irrelevance within the framework of the Habsburg Empire, external factors began to act on its internal constitution, exacerbating local strains by adding the tensions of external political and economic competition. In the north, the German petty states yielded increasingly to the dynamism of Prussia, striving to forge a new German *Reich*. In the south, the rise of a united Italy challenged Habsburg preeminence and Tyrolese security.

ITALIAN UNIFICATION

In Italy, the victories of the Napoleonic armies in the wake of the French Revolution unlocked nationalist aspirations, which discharged their tensions against the Austrian polity that dominated the northern Italian mountains and plains. Industrial development, together with the successful commercialization of agriculture, had created an active economic and poltical pivot in the Italian north even before 1848. The Kingdom of Sardinia, with its capital at Turin in Piedmont, and since 1815 in possession of the port of Genoa, became the chief protagonist in the struggle to oust the Austrians. Prosperous Lombardy, with its thriving center at Milan, was wrested from the Austrian power in 1859; Venice was yielded in 1866. The *Risorgimento,* as the movement for Italian unity came to be known, based itself on an alliance of Piedmontese expansionists, liberal nobles, rationalizing entrepreneurs, and bureaucrats, led by Camillo Cavour and the mushrooming middle sectors, comprising professionals, teachers and students, craftsmen, capitalist farmers, and others who dreamed of republican institutions and followed Giuseppe Mazzini and Giuseppe Garibaldi. Although inspired by the ideals of the French revolution, the republicans placed the question of national unity above that of social reform, and in time came to fear social revolution. Their energies were directed outward, toward the redemption of the Italian lands still groaning under foreign yoke. Any political alliance-building within Italy came to depend on a politics of military opposition to the Austrian polity. Therefore, the Tyrolese increasingly found themselves charged with the defense of the empire on its Italian front. In 1848, for instance, 144 companies of Tyrolese sharpshooters were called out to guard the southern perimeter alongside regular units of the Austrian army. The call to arms was repeated in 1859, during the campaign in Lombardy. In 1866, 40,000 sharpshooters and militiamen took the field against Italy. The loss of Lombardy and Venice drew the frontier along the boundaries of the Italian-speaking part of the Tyrol, the Trentino; the Trentini and the inhabitants of the so-called Littoral, with Trieste at its center, became the last Italians within Austria.

GERMAN UNIFICATION

At the same time, the Tyrol began to experience the pull of an ever more unified and stronger Germany, especially after Austria was forced to withdraw from the German Federation in 1866, as a result of her defeat by Prussia. Under Prussian hegemony, German industrialization advanced relentlessly, spurred by an expansionist militarism. The vacillating German bourgeoisie, in turn, yielded to the direction of the Prussian bureaucracy and army. The famous *Sammlungspolitik*—a politics of political consolidation—created a liberal–conservative alliance. The alliance was used first against the political representatives of particular regions and later against the labor movement led by the Social Democrats at home and competitive nations abroad. The capstone to the alliance was placed in 1879, with an agreement between industrialists and agrarians that bought agrarian support for an expansionist industrial policy at the price of protective tariffs for grain at home. The political expression of the alliance was the Pan-German League, created in 1890, which brought together liberals, conservatives, anti-Semites, and agrarians in support of the expansionist program. This organization, in turn, threw its support to the multitude of nationalist organizations, such as the Colonial League and the Society for Germans Abroad.

Sentiments favoring a politics that would maintain Austria as a buffer state between Germany proper and the growing threat posed by Slavic demands for liberty and independence increasingly gave way to a *Grossdeutsch* ideology, which foresaw the unification of all German speakers within a single German *Reich*. Within the Habsburg monarchy this sentiment grew in proportion to the realization by German speakers that they were but one minority among other minorities within the Empire. It was fed by strongly anti-Semitic currents, which saw in economic development and the dislocations associated with capitalism the hand of Jewish bankers and entrepreneurs. Anti-Semitism went hand in hand with anti-clerical tendencies directed against a church that supported the supranational policies of the dynasty, and could thus be accused of backing the claims of the Slavic and Italian minorities. Much German nationalist fervor was directed against the Czechs, whose successful economic expansion especially infuriated the German-speaking middle classes.Yet it also found an echo in the Tyrol, where anti-Italian sentiments could easily be tapped for a more aggressive nationalist pan-Germanism. Whereas the Tyrolese remained largely loyal to Church and dynasty, they could not but respond to a political propaganda that boosted their status as German speakers in a sea of culturally alien ethnic groups. Increasingly, the responses of the Tyrolese were determined not only by their own internal situation within the Habsburg Empire, but also by the powerful pincer movement from the north and south, by the convergent pressures of Italian *irredenta* and pan-Germanism.

NATIONALIST RIVALRY

The effect of these convergent pressures has been to turn the cisalpine Tyrol, partly German speaking and partly Italian speaking, into an ethnic battleground. In the wars against the Napoleonic incursions and during the Tyrolese *levée en masse* against foreign rule under Andreas Hofer, German-speaking Tyrolese and Italian-speaking Trentini still fought side by side (Hirn 1909, 387; fn. 3, 667; Zieger 1926: 158). But the rise of the *irredenta* slowly increased the social distance between the two populations. The development of this widening schism was gradual, and its consequences anything but dramatic. There was some conspiratorial activity early in the nineteenth century: a cell of the underground society of the Carbonari was organized in Trento in 1813, and when Mazzini's Young Italy Society was joined by some Trentini in 1831, the Trentini were promptly arrested by the Austrian police. There were some efforts at invasion from the outside: an Italian free corps invaded sections of the Trentino in 1848, and Garibaldi led an incursion almost to the gates of Trento in 1866. Yet:

> As far as one can tell, articulate Italian sentiment was confined to the cities, and there is no good reason to believe that the anti-Austrian agitators—the 'italianissimi,' as they were dubbed—were more than a small minority or that they carried any serious weight even in the cities. The informers of the Viennese government in the region during the period between 1815 and 1848 consistently reported that the population was peaceful in a national and political sense. But the idea that the old order was not to be regarded as natural or unquestionable had been introduced, and the common language and culture of the region formed a network of communication that facilitated its cirulation [Greenfield 1967: 497].

This remained true throughout the century, in the sense that Italian political demands in the Trentino were oriented toward the achievement of some measure of autonomy within the Habsburg Empire, rather than toward total severance from it.

With the passage of time, the Liberal Nationalists—the nationalist, irredentist, antisocialist, monarchist, and anticlerical party, with its greatest strength among the leading burghers of the bigger Italian towns in the Trentino—began to lose out to a Christian Socialist Party, the *Partito Popolare*. Contrary to the Liberal Nationalists, whose stand was ideological at the expense of any interest in economic issues, the *Partito Popolare*, led in the main by the rural Italian clergy, was oriented toward the rural population and its matter-of-fact interests. The Italian peasantry of the Trentino was not interested in nationalist politics; its political concerns were primarily those of valley and village, and it favored solutions to economic questions above any movement toward national liberation. The clerical leadership responded to villagers' interests with the creation of a network of rural credit institutions and cooperatives. At first, the clerics in charge were rather hostile to the new Italian state, which they regarded as the work of anticlerical and antiecclesiastical Freemasons; later, they began to use irredentist slogans themselves, more to gain leverage on the political processes within the Habsburg polity than for any reason of outright annexation by Italy (see Kramer 1954; Monteleone 1971: 30–32). In this they were joined by many officials, clerks, and craftsmen of the towns.

Nineteen hundred also saw the rise of a small socialist party, which gained adherents among the workers of Trento, Rovereto, and Riva, and among the Italian-speaking migratory workers from the Trentino holding temporary jobs within the Tyrol and Vorarlberg. The socialists also gained a small measure of influence in the countryside through the organization of a peasant league (*Lega dei contadini*). Strongly in support of the socialist party were university students who had no choice but to attend either the German-speaking university at Innsbruck or the Italian universities in the Kingdom of Italy, and whose chances for political mobility were hampered by the relatively small size of the Italian minority within the Habsburg Empire (Kramer 1954; Salvemini, in Gatterer 1968: 60). There was, however, a significant increase in Italian organizations "which declared a purpose of broader appeal, namely, the tutelage and diffusion of Italian culture [Greenfield 1967: 510]." The first of these was the *Pro Patria*, founded in 1885 in Rovereto, but declared illegal in 1890. Of more enduring importance was the Dante Alighieri Society, founded in Rome in 1889, which lent its support to Italian schools, newspapers, and intellectual circles, and which promoted Italian culture and identity among Italian migrants abroad. Also significant within the confines of the Trentino was the growth and development of the Italian Alpine Society as a competitor to the nationalist German and Austrian Alpine Association.

Among the German-speaking Tyrolese, too, the largest political organization was also a Christian Socialist Party, like its Italian counterpart, oriented to the problems of the rural population. Strongly proclerical and prodynastic, in the traditional Tyrolese sense, it was also interested in the maintenance of the rural social order and in the development of Christian corporate groupings along supposedly medieval lines. At the same time the Christian Socialists favored continued German dominance in the Tyrol, saw Italy as a haven for atheists and Freemasons, a state led by a godless king and defended by an army "without chaplains [Gatterer 1968: 126]." In its political attitude toward Italian aspirations, the party was "primitively anti-Italian [Gatterer 1968: 179]." The Socialists, the only party to favor autonomy for the Trentino, were weak. A German National Party was strong in the towns, but not in the rural hinterland. Yet pan-German sectors were increasingly vocal and demanding. Just as the Italians organized their Dante Alighieri Society, so the Germans developed a panoply of increasingly aggressive, German cultural societies. A *Deutscher Schulverein* was founded in 1880 to further German schools in the German-speaking islands within the Trentino. The pan-German *Allgemeiner Schulverein* of Berlin entered the scene, as did the *Schulverein Südmark*, founded in Graz in 1889 and by 1894 active in the Tyrol, where it combined an interest in German schooling with attempts to further German colonization in the Trentino through purchase of land. A *Tiroler Volksbund*, under clerical leadership, arose in Innsbruck in 1905, and began to dream of "re-Germanizing" the Trentino. All these groupings, were under the strong influence of the Pan-German Association, founded by the twin forerunners of Austrian National Socialism, Georg von Schönerer and Karl Herrmann Wolf, who soon became the idols of the Austrian university students in Innsbruck. Too weak politically to gain representation in parliament, they were, nevertheless, of great intellectual importance in orienting the Tyrolese toward a more aggressive German nationalism.

As these attitudes grew sharper, and as the protagonists began to challenge each other—over the erection of a monument to Dante Alighieri in Trento in 1896, over the dedication of a monument to the German troubadour, Walter von der Vogelweide, in Bozen in 1889, over nationalist slogans printed on match books, in riots between German and Italian students at Innsbruck in 1904—political options that had been open in the nineteenth century disappeared with the passage of time. The request for autonomy for the Trentino was never granted; the politicians of the dynasty feared correctly that granting autonomy to one minority would unleash similar requests elsewhere. By 1905, Cesare Battisti, the leader of the Trentine socialists, had decided the nationality problem could be solved only through the destruction of the monarchy; and in these sentiments he was increasingly joined, though on a more covert level, by the Archbishop of Trento, Celestino Endrizzi.

WORLD WAR I

The outbreak of war in 1914 furnished the opportunity for a radical solution of nationalist aspirations. In 1915, Italy declared war on the side of the Allies against the Central Powers. Many Trentini fought on the Austrian side, even against their ethnic brothers in the Kingdom of Italy, but 2000 Trentini joined the Italian army to fight for Italy. With German and Austrian defeat, the Italians not only laid claim to the Italian-speaking portion of Tyrol, but asserted the right to occupy the entire German-speaking area up to the Brenner. Whereas the socialists and even a large section of the *Partito Popolare* opposed annexation of the German-speaking section, the imperialist *irredenta* had its way. The Tyrolese sought various ways to retain the cisalpine part of the territory: for a brief moment they contemplated a Tyrolese declaration of independence. Some representatives of the Tyrolese Diet are said to have approached the King of Italy in order to urge inclusion of the entire Tyrol within the Italian Kingdom on the grounds that they were Tyrolese long before they were Austrians. The offer was rejected (Toscano 1967: 36–43). Thereupon they put their hopes in annexation to Germany, an option that was over-whelmingly approved in the plebiscite of 1921. This move, however, was vetoed by the Allies. The South Tyrol became Italian by the peace treaty of St. Germain in 1919, although the Tyrolese members of the Austrian delegation to the peace conference refused to sign the treaty.

Abandoned by Austria, the South Tyrolese prepared for an initial accommodation with Italy through the organization of a *Deutscher Volksbund* (German People's League), which would include both the Christian Socialists, strong among the rural population, and the pan-German political representatives of the Bozen burghers. For a brief time it seemed as if the development of a viable parliamentary repre-sentation would allow the German speakers within the new Italian province to enter into the wider Italian political process. This hope, however, quickly fell victim to rise of Fascism. The war had shaken Italian society to its roots; a socialist seizure of power, based on an alliance of industrial workers and agricultural laborers in the northern plain seemed imminent. Fascism capitalized on middle-class fears of a violent takeover; it permitted the industrial and the agrarian elites to maintain their power through an alliance with elements of the middle class and dissatisfied war veterans, who could be mobilized in political violence against socialist aspirations (Germani 1972; Gramsci, in Cammett 1967: 178–179; Sarti 1971). Throughout 1920, the Fascist *squadre* devastated Emilia, Tuscany, and the Po Valley, effectively breaking up peasant organizations and disorganizing the opponents of a Fascist seizure of power. By 1921, they were active in the new Italian Tyrol. On March 6, 1921, Black Shirts and members of the German People's League clashed at Neumarkt/Egna; on April 24, Black Shirts under Achille Starace, later to be the secretary of the Fascist Party, attacked a street crowd gathered for the fair at Bozen/Bolzano. On October 1, 1922, the Fascists seized control of Bozen and demanded the resignation of the German-speaking mayor, the introduction of Italian into the schools and government offices, and bilingualism in all public signs and notices. On October 3–4, they seized control of Trento. The Trentine Fascists had shown a willingness to make business deals with individual, German-speaking Tyrolese. They intended the economic integration of the Province of Bozen into their domain at Trento, without at the same time carrying out a wholesale policy of denationalization. Therefore, the Fascist ideologues pushed aside their regional representatives and put the affairs of the South Tyrol directly under the control of central headquarters. With Mussolini's successful March on Rome, October 26–30, 1922, the Fascists seized power and put an end to any thoughts of Tyrolese autonomy within the Italian state.

FASCISM

Fascism underwrote a policy of denationalization for the German minority, much as it did for the French speakers in the Val d'Aosta to the west and for the Slavs in the southern Slavic borderlands. The very name "South Tyrol" was prohibited; the South Tyrol became Alto Adige. At first Alto Adige was incorporated into a larger region of Venezia Tridentina, with its seat at Trento; but soon after it was given separate status as the Province of Bolzano and was placed under direct control from Rome. The traditional Tyrolese elective system gave way to rule by an appointed provincial prefect who, in turn, appointed or removed at will local mayors and communal secretaries. Many old communities, deemed too small to support the

new bureaucratic apparatus, were incorporated into larger administrative units. Italian became the official language, both for administration and in the courts. Officials had to be Italian speaking, to have
attended Italian schools, or to have served at least three years in the
Italian bureaucracy. Communities received Italian names. All public
inscriptions were to be in Italian, including inscriptions on tomb-
stones. Family names were to be Italianized, though the operation
took longer than expected and never got beyond the letter B. Italian
became the obligatory language in the schools from the first grade
on; the use of German was prohibited. This was also true of religious
instruction in the schools. Similarly, Mussolini forbade the lighting of
Heart of Jesus fires, symbolic of the unity of Tyrol, on all mountain
tops on the third Friday after Pentecost.

The imposition of administrative and educational controls was
paralleled by economic sanctions. German banks were put under
Italian control. The long-standing Tyrolese legislation that guaranteed
impartible inheritance in land was abrogated in 1929: nearly half the
homesteads were to feel the impact of the Italian pattern of partibility
between 1929 and 1954 (Leidlmair, 1958: 136). An organization—the
Ente Nazionale per le Tre Venezie—was created to buy up land from
German speakers and transfer ownership to Italians. We shall explore some of the local phenomena associated with these changes in
ownership and inheritance patterns in Chapter IX. Industrialization
of the area was initiated under Italian auspices, beginning in 1924
with the development of hydroelectric works and culminating after
1935 in the construction of a major complex of heavy industry in
the Bolzano/Bozen area, comprising steel mills, aluminum works,
magnesium works, and an automobile factory. This industrial complex was manned almost entirely by Italians. Italian labor was drawn
into the province through government propaganda and promises,
and intermarriage with German speakers was encouraged and even
subsidized by the provincial government. Efforts of the German-
speaking South Tyrolese to call attention to their growing distress
came to nothing. The Fascists put down all attempts at legal representation or expression within Italy, and a defeated and impoverished
Austria or Germany lacked the necessary political leverage on the
international level to create external pressures against Mussolini's
regime.

At first the opposition to Fascist-imposed Italianization was
cultural, and only secondarily political. Deprived of official representatives, the South Tyrolese found their main defenders in the rural
clergy, exemplified most especially by Canonicus Michael Gamper,

who strove with might and main to counter Italian influence and pressure by continuing to assert the "natural right" of the priesthood to preach in German and to continue to impart a German education both in clandestine "catacomb" schools and through clerically oriented newspapers. Yet, gradually, the rise of National Socialism in both Germany and Austria kindled more desperate hopes, especially among the burghers, the marginal peasantry, and the young, for a radical solution that might someday bring the South Tyrol into a Greater Germany. Support for this hope came from Austrian and German citizens who continued to live in the South Tyrol, and among whom a Nazi party was organized in 1932. Further support was lent by the Popular League for Germanism Abroad, an offspring of the Viennese German Educational Association (*Deutscher Schulverein*), which supported physically and financially the educational work of the "catacomb" schools. A third, and not insignificant, source of support were the adherents of the budding Nazi movement in Austria, especially in the Tyrol; among them, even in years to come, dreams of Tyrolese unification would often take precedence over unquestioning obedience to the dictates of the National Socialist *Führer*. The shift toward a more radical political solution in the South Tyrol itself is best exemplified by the growth of an illegal youth organization in 1933, led by Peter Hofer, a Bozen tailor. By 1939 it counted 1500–2000 members. It had its beginnings in the Catholic youth movement, in which Hofer had played a prominent part. Yet with the passage of time, the organization moved increasingly toward an accommodation with National Socialism, until Hofer himself became an obedient instrument of the *Reichsführer-S.S.*, Heinrich Himmler.

THE PLEBISCITE

The German *Führer*, Adolf Hitler himself, took a different line from the very first. In speeches from 1922 on, in a special publication entitled *Die Südtiroler Frage und das Deutsche Bündnisproblem* (The South Tyrol Question and the Problem of Germany's Alliances), which appeared in 1926, and in successive editions of *Mein Kampf*, he resolutely declared his willingness to sacrifice the aspirations of the South Tyrolese to his larger designs for an alliance between Fascist Italy and National Socialist Germany. The South Tyrol was not going to stand in the way of his "wave of the future." In 1936 the Popular League was put under Party supervision and called on to cease all activities on the behalf of the South Tyrolese. Similarly, the hopes of Tyrolese and Bavarian Nazis that the *Anschluss* with Austria might also bring about the deliverance of the South Tyrol, were icily quashed. Instead, Mussolini and Hitler, in 1939, set about cementing their "brutal friendship" by solving the South Tyrol problem once and for all through relocation of the troublesome population in the *Reich*. In a plebiscite, German-speaking South Tyrolese were to be forced to opt for one of two fateful alternatives: either vote to stay in Italy and accept Italianization, or retain German identity and move to Germany. Various leaders of the S.S. dreamed of relocating the South Tyrolese in Burgundy, which would thus regain its ancient Germanic character, or in eastern Europe. In 1942, Hitler commented with delight on a scheme to transport the South Tyrolese to the Crimea, once a Gothic stronghold very favorable to the maintenance of ethnic groups, and "a land of milk and honey," vastly more productive than their mountainous homeland. He stated:

> The transport of South Tyrolese to the Crimea offers no special physical or psychological difficulties. They need only make the voyage down a German stream, the Danube, and they'd be right there [Picker 1951: 155].

The plebiscite, or *Option*, was held in December 1939. Of 267,265 participating voters in the four provinces of Bolzano, Trento, Belluno, and Udine, 185,365 voted to leave for Germany, 38,274 opted to stay in Italy, and 43,626 refrained from voting. Those voting to leave Italy included more than half the Ladinsh-speaking population, whom the Italians had always classified as ethnic Italians (Huter 1965b: 341). Those opting for Germany included a majority of the peasantry and a fifth of the clergy. Those choosing to stay in Italy comprised the larger part of the Bozen/Bolzano bourgeoisie, the landed aristocracy, 80% of the clergy, and a minority of the peasantry (Rusinow 1969: 260).

    The plebiscite itself unfolded in the most tense atmosphere of political pressure and counterpressure. The National Socialists developed a masterful and harassing propaganda that equated the retention of German identity with a "return" to the *Reich*. Many people who were not Nazi opted for Germany in the end as a protest against the Fascist Italian state. Others feared a repressive policy of further Italianization, a fear fed by rumors of eventual relocation of the remaining South Tyrolese in other parts of Italy, or even in Ethiopia. Some voted for Germany in the hope of an eventual return to the South Tyrol under the aegis of a victorious postwar German *Reich*.

Still others undoubtedly voted for Germany in the hope of participating in Greater Germany's prosperity and power, then at its zenith. Some were outright Nazis. Economic considerations undoubtedly played a part in the decision. Italian competition, together with the Great Depression of the 1930s, had weakened the German-speaking peasantry: their credit institutions were under Italian auspices; their markets had been curtailed; they were faced with a growing mountain of debts. Tourism, especially German tourism, had declined markedly, prompted by German restrictions on monetary exports. Artisans and small merchants had also been hard hit. Thousands of children could not find public employment nor work in the new Italian industrial complex of the area. The motivations of those who voted for Italy were similarly mixed. Many of those who voted to stay in Italy were not, by virtue of that fact, pro-Italian; they were willing to continue some kind of accommodation with the Fascists to retain a foothold in their ancient homeland. Others, including most clergymen, were opposed as Catholics to the godless Nazis and felt that remaining in Italy also guaranteed the continuity of their South Tyrolese Catholicism. The conflict between *Optanten,* those voting to depart for Germany, and the *Dableiber,* the stay-at-homes, divided whole villages and families; the hatreds, recriminations, and mutual suspicions engendered by the plebiscite have remained alive to this day.

   Although the vote had gone overwhelmingly in favor of relocation into the *Reich,* relocation was realized only in part. Some 75,000 departed between 1939 and 1945, the last year of World War II; another 20,000 were called up to serve in the German *Wehrmacht.* The highest incidence of departures occurred between the beginning of 1940 and the fall of 1941; the highest percentages of departures occurred in occupational categories without direct roots to landed property. Most of those who left were artisans, laborers, workers in transportation or in domestic service, or involved in the restaurant or inn business. By 1943, in contrast, 91% of the population engaged in agriculture had remained; only 9% had emigrated.

   A variety of factors worked to retard the peasant migration. Legal proceedings for monetary compensation, guaranteed by the plebiscite agreement, took time and were often delayed by German officials who hoped for a change in the situation, or by Italian officials who balked at the high costs. Changes in German political fortunes also lessened the enthusiasm for transfer to Germany. Actual contact with the blessings of National Socialism brought about a change of mind: migrants to Germany began to give vent to criticism;

some were confined in concentration camps. Some migrants declared they had been deceived by false promises into voting for Germany; clandestine returns to Italy became common (Rusinow 1969: 261–262). Some landowners and South Tyrolese, on the other hand, charged with overseeing the migration were not unwilling to further the emigration of their poorer or more difficult neighbors, while holding to the hope that such ostensible cooperation with the *Führer's* plans would allow them to delay their own departures (Gatterer 1968: 620). The final *coup de grace* to resettlement, however, was imparted by the events of 1943, when Italy collapsed under the impact of the Allied invasion, and Germany seized control of the Prealps—Bolzano, Trento, Belluno—in an attempt to secure the Italian north for the Axis Powers. Just as the plebiscite of 1939 had issued in the ironic result of a vote to "return" to the *Reich* to which none of the voters had ever belonged, so the Italian collapse for a brief period recreated the unity of the South and North Tyrol, but under the aegis of a Germany clearly in military decline. To many South Tyrolese, the German occupation meant that all their dreams of reunification with their brothers to the north had finally come true; but an Allied victory over Germany would once again underwrite a resurgence of Italian claims, since the Kingdom of Italy was again in the Allied camp.

On the surface, the Germans maintained a distinction between the zone of the Prealps and the *Reich*, but the *Gauleiter* of the Tyrol, Franz Hofer, like many other *Gauleiters* in the overexpanded *Reich*, managed the prealpine provinces very much as part of a semi-autonomous fief of his own (Rusinow 1969: 298; Toscano 1967). He did not always have his way, as when he had to yield in his intention to fuse South Tyrolese military organizations with his Tyrolese sharpshooters' organization; the S.S. instead grouped them systematically into S.S. regiments, supplementary battalions, and S.S. police organizations. But he managed to forbid all political party activity in the South Tyrol, whether Nazi or Fascist. In the Trentino he appointed as prefect a prototypical Liberal National of the pre-World War I period, Alberto de Bertolini, rather than a Fascist; in Bozen Province he was able to replace as prefect the S.S.-oriented Peter Hofer, who died in an air raid in 1943, with Karl Tinzl, a former Austrian official, notable in the post-World War I German *Verband* and a non-Nazi.

Though the German element was clearly dominant in the South Tyrol, maintenance of an Italian–German bilingualism in public affairs seemed to reestablish the status quo prior to World War I. At the same time *Gauleiter* Hofer resolutely backed the efforts of the S.S. to stamp out various resistance groups that had begun to rear their heads in the area. In the Trentino and Belluno, groups of Italian anti-Fascists took to the hills, but their center in Trento was wiped out, as was the Italian National Liberation Committee in Bozen. The same fate befell a German-speaking partisan group, mostly concentrated in Passeier; but an Andreas Hofer association with pro-Austrian and autonomist tendencies survived. As Germany neared defeat, Hofer made ever stronger attempts to secure the continued unity and integrity of the Tyrol, even through clandestine contacts with the Allies. But these efforts proved in vain. On April 29, 1945, the German army in Italy capitulated, and American troops, followed by Italian army units, occupied the Prealps. On May 3, Italian police troops once again raised the Italian flag on the Brenner.

CHAPTER **IV**

The Economic Development of the Rural Sector

We have traced the political trajectory of the Tyrol and we have followed the ups and downs of its economic development. The resulting political and economic drama is played out differently, however, on the level of the larger polity and on the level of the village. How, then, did these political and economic forces affect the Tyrolese and Trentine peasant, striving to wrest a livelihood from a difficult and often forbidding land? It is the peasant who must ultimately square the demands of his environment with the demands of the outside world. Whether he is interested in mere survival or in economic gain, he must bring his technological and social organization into concordance with the physical environment. Yet, at the very same time, he must answer to the promptings of the outside world.

No matter how remote a mountain village, it does not exist in isolation. It maintains contact with other communities and develops linkages to a wider polity and economy: the growth of a Tyrolese political economy was felt most keenly in the lowlands, but exerted its influence even in remote upland villages. Yet the region is no more a self-contained world than is the mountain village, and at least since Roman times economic ties have bound the Tyrol to other parts of Europe. The strength of these ties has waxed and waned, and their nature has changed over the years; their effects on peasant communities and growing towns thus has not been constant. Still, we believe we can identify long-run trends in the economic life of the Tyrolese and Trentine peasant. One of these trends is manifest in the peasant's continual gain in freedom to manage his own affairs, to cast off dependency on the feudal overlord's estate to become lord of his own lands—an independent actor in politics and economics. Another trend has led the peasant to forego the economic self-sufficiency afforded by subsistence production for a commitment to the market. In the following pages we will discuss these trends at length.

Although events in the Tyrol parallel those that accompanied the transformation from feudalism to capitalist nation-states in other parts of Europe, the ecology and history of the region produced a unique expression of this common European theme (see Dopsch 1930; Tremel 1969; Wopfner 1951–1960). Again, the geomorphology of the Tyrolese Alps has played an important role in economic development. It has influenced both the nature of agriculture and the development of commerce and industry. The original settlement of farming folk in the Neolithic produced a distinctive variant of European mixed farming, and each subsequent stage in the development of the Tyrolese economy has been influenced by previous modes of adaptation.

The challenge of the environment to people intent upon farming in the mountains is to find sufficient land suitable for cultivation and to develop other productive uses for land that cannot be brought under plow. As early as the Neolithic Age, a pattern of land use was practiced in the Alps that assigned the most favorably situated areas to plowland and tended meadows, relegated somewhat less productive land to secondary meadows, pasture, and forest, and used the least favorable areas only for the harvest of wild flora and fauna, or left it as wasteland (Higham 1967). However, the evaluation of land is to a degree relative to the needs of the farmer, so that what may be relegated to secondary uses in one era may be valued as prime land in the next. As population pressures increase, and as the commercial penetration of the Alps has proceeded apace, there has been ongoing reevaluation of the use to which any given area of land is put.

The development of the commercial sector of the Tyrolese economy, too, was constrained by the form of the land. Trade routes could follow only the courses of major rivers, those leading to the most accessible passes—the Brenner and the Reschen being the most heavily traveled. As we have seen in Chapter II, the very shape taken by the *Land Tirol* was determined by the desirability of controlling the passes and their access routes. Commercial, military, and administrative posts grew up along these routes, those located at the junctures of trade routes becoming especially prosperous. Thus, Innsbruck, long the largest city in the Tyrol, is located at the intersection of the main north–south and east–west routes, whereas Bozen, the largest city south of the Brenner, lies at the point where the northbound traveler must decide whether to follow the route of Brenner or Reschen.

The flow of goods between points within the Tyrol also takes advantage of the most convenient routes. Movement of goods between communities located in the same valley is easy, but mountains inhibit such movement between different valleys. Thus, the movement of goods, whether stimulated by feudal rents or market profits, has been from mountainside to mountain valley, from the high valleys downstream to major valleys, and hence along the valley floor to their final destination. A map of the flow of goods would show main arteries and tributary veins, rather than the gridiron pattern characteristic of lowland transportation systems. Moreover, country folk in the Tyrol have become involved in commercial development in direct proportion to their proximity to new towns and trade routes. Whereas political and economic ties between administrative and commercial centers and villages lying near them have been tightly drawn, ties with villages farther away have been much more tenuous.

We find this constellation of characteristics already unfolding as history dawns on the Tyrol, conquered by Roman legions. The Empire was attracted to the land of the Rhaeti for strategic reasons, in order to defend its northern frontier; the Roman garrison and administrative centers, however, were set up along the main arteries of communication and attracted merchants who sought to move goods both to these centers and to destinations beyond the mountains. Roman villas were built, but remained few in number and confined to the main valleys. These settlements and movements of goods drew local people into service. The Rhaeti contributed few goods to the long-distance trade and drew little from it for local consumption: rather, they formed a labor pool of guides and porters for alpine trade. No doubt the immediate hinterland supported the Roman centers with agricultural produce, but the population aggregate requiring support was small, and the effects on productive techniques and strategies were correspondingly insignificant. More distant valleys continued to produce only for themselves: many, like the Anaunia, remained beyond the reach of Roman administrative control.

EARLY MIDDLE AGES: SIXTH–ELEVENTH CENTURIES

The economic organization of the Tyrol in Roman times anticipates certain modern trends, but their development has not been continuous over the entire period of Tyrolese history. In the early Middle Ages, following upon the collapse of Rome, the political and administrative organization that had served it was gradually withdrawn. At the same time, long-distance trade withered away and old trade routes fell into disuse. The Germanic settlers who moved into the Tyrol from the north and northwest in the seventh and eighth centuries used these same routes, which from time to time were crossed by armies bent on conquest. However, transportation of goods destined for distant markets ceased. Production in the countryside, no longer geared to the requirements of an empire, came instead to support a network of local secular and ecclesiastic nobility.

In the feudal economy emphasis was on local production for local use. Each peasant household strove to provide for the full range of its perceived needs from its own land and labor, and for the execution of more specialized tasks sought assistance from those in the locality who had the necessary skills. Within the village, each household therefore produced about the same sorts of goods and services as every other. Similarly, moving from village to village, or from manor to manor, there was little variation in the kinds of crops grown or animals raised. Economic interdependence was thus minimal, and following the decline of the Roman towns it was no longer possible to market produce to a nonrural population.

The struggle of the Tyrolese peasant to satisfy all his needs from his own resources was paralleled by the nobleman's dependence on the production of the peasants over whom he held domain. Since rents were paid in kind, and since opportunities to convert produce into other commodities were as restricted for the lord as for the peasant, the life styles of the nobility differed little from those of their subjects. In the absence of a market where local produce could be converted into some other form of wealth, there was little incentive for the lord to stimulate productivity by increasing peasant rents. His household could consume only so much sauerkraut, smoked bacon, and bread. Even the temptation to increase the size of the household by adding retainers must have been slight: more dependents would increase the administrative burden without providing much compensation in the form of benefits to the lord.

Nevertheless, there was tension in the relationship between lord and man, a tension centering on the disposition of the fruits of peasant labor. Any increase in the goods and services exacted by the lord would be regarded by the peasant as an infringement on his traditional rights; conversely, the lord would consider the withholding of peasant rents to be a violation of tenurial agreements. Moreover, a reduction in productivity, whatever the cause, placed a strain on the relationship, for both lord and man would seek to maintain their customary rate of consumption despite a smaller harvest. In this delicate balance, the lord gained greatest control over the produce of his domain when he exercised direct management of both production and distribution of the harvest. Having both detailed knowledge of what was produced and familiarity with the serfs' needs, he was able to allocate the produce of his lands to satisfy their minimum requirements, claiming the remainder for his own storehouses. On the other hand, the peasant was best able to direct surplus produce toward his own ends when the noble exercised little supervision, and the peasant himself managed his own estate.

In the early Middle Ages the relationship of lord to peasant favored management by an elite. Bavarian nobles who moved into the Tyrol in the seventh and eighth centuries brought with them the feudal arrangements of Bavaria. These feudal rights were not only applied to the serfs who migrated with them, but they were also extended to the original inhabitants of the Tyrol who happened to reside within their domain. The majority of these serfs were landless families who lived and worked on the lord's estate; a minority worked their own shares of land but paid rent to the lord and furnished labor for his estates. Thus, the domain of any lord was likely to include both land divided into fields worked by peasants on their own behalf and undivided sections worked by a combination of labor dues. In either case, the nobility exercised close control over production and made or approved important decisions on the distribution of the harvest.

If close managerial supervision of the domain allowed the lord maximum control over peasant production, it also contributed to a low level of political centralization. Since virtually all the wealth of society derived from peasant labor, political hierarchies could develop only to the extent that central elites were able to mobilize a portion of the rents collected from the peasantry. Local nobles, however, were not representatives of a centralized state; each noble held domain in his own right. Thus, such dues and services as might be rendered to a prince, such as the Bishops of Brixen, Chur, and Trent, were not taxes collected by him as a servant of the state, but an obligation placed on the lord himself. Local and territorial rulers were therefore in competition for peasant surpluses, which produced a tension between nobles analogous to the tension between lord and man. The local nobility also preferred to keep the overlord at a long arm's length.

The degree to which the territorial elite was willing to assert its claims in competing with manorial lords for the peasant's produce depended on whether the administrative effort required to control these local nobles was compensated by a level of economic returns sufficient to make the effort worthwhile. The domains lying within easy reach of regional centers might be effectively controlled, but in dealing with more distinct domains, the prince had to consider problems of moving over Alpine terrain and of dealing with nobles well entrenched in fortresses that made the most of the natural advantages of the landscape.

Castles are thick on the ground in the Tyrol and bear witness to the political and economic fragmentation of the times. Each lord held domain over his lands and from his fortress home both defended it against encroachment and plotted the expansion of his domain at the expense of neighboring lords. However, during the early Middle Ages neither political intrigue nor military conquest were effective in carving out larger political units. Local elites retained a high level of autonomy as long as the Tyrolese economy continued to be based on subsistence agriculture. Despite the conservative outlook of both peasant and noble in the Tyrol, new variables beyond their control were being introduced into the economic equation by the eleventh century. Population increase, transalpine trade, and the advent of a mining industry in the Tyrol were accompanied by the introduction of capitalist economic relations into the Tyrolese countryside. Politically, these economic trends were accompanied by an erosion of the strength of the land-based nobility, the growth of political centralization, and the freeing of the peasantry from feudal obligations.

LATE MIDDLE AGES: TWELFTH–SIXTEENTH CENTURIES

THE DEMOGRAPHIC PROBLEM

By the eleventh and twelfth centuries, population increase in the Tyrol began to strain available resources and to create problems for the feudal order. The niche occupied by peasant farming communities along the terraces in the main river valleys had absorbed as much population as it was capable of supporting comfortably. As the population continued to increase, tensions increased between peasant and lord over distribution of the harvest. Without an increase in agricultural production, the peasantry as a whole would require an ever greater proportion of the total harvest as their numbers increased. In this case, the nobility had to accept a reduction in rents. On the other hand, if feudal dues were to remain at a level satisfactory to the lord, the peasant would have to tighten his belt—an unsatisfactory solution from his point of view.

Both lord and peasant could agree on the desirability of limiting the number of families with access to land. This could be done by legally preventing the division of an estate and by reserving its harvest for a single domestic unit. By so limiting the number of families with access to land, the lord protected his rents against an expanding population. At the same time, the peasants on these estates were guaranteed the use of resources sufficient to meet subsistence needs. In time these arrangements were codified, the holdings made legally indivisible, and the geschlossene Hof, as this type of impartible estate is called, became the dominant tenurial arrangement in areas of German settlement in the Tyrol. The peasant obtained inheritable rights to the estate in return for the payment of a hereditary rent to the lord. The franchise was passed on through lineal descent, preferably through the first-born son. While the new tenurial arrangement protected the productivity of the land against the expanding population and thereby protected the rents and dues that supported the rural elite, it did so only at the expense of the lords' managerial control. The peasant now managed his own production and controlled the disposition of his own harvest. Although he was still faced with payments of rent, the peasant gained much economic freedom. The nobility had sacrificed direct control over peasant harvests in exchange for a guaranteed, fixed rent.

The establishment of indivisible estates insured that the land would continue to produce a surplus that could satisfactorily underwrite the life styles of both lord and peasant, but did not bring about a final solution to the population problem. This could be accomplished only if those denied a share in the estate's produce were provided alternative means of support. Monastic orders were already functioning in the Tyrol, and as in much of Europe, they had served for years to absorb many of those without prospects. The founding of new monasteries after the tenth century, many high in the mountains above the altitude of permanent settlement, may well have been a response to population growth. Whatever prompted their proliferation, monasteries did serve as the vanguard for an expansion of the land under cultivation to higher altitudes. By the twelfth century, high valleys and plateaus overhanging the main valleys were being converted from seasonal occupation by herdsmen to year-round settlement by agricultural communities. Sponsored by both secular and ecclesiastic elites, as well as monastic orders, the ongoing settlement of this frontier was for a time able both to absorb the growing numbers of peasants and to serve as domain for the landless members of an expanding nobility.

Yet the very existence of this frontier, although helping to ameliorate the problem of surplus population growth in the low valleys, served to undermine the very feudal relations that it appeared at first glance to sustain. In Chapter II we noted the correlation between the opening of this frontier and the trend toward a peasantry free from feudal obligations. Here we take up the economic basis for this trend.

The communities being settled at that time were on lands both less productive and less accessible than those at lower altitudes. Land being less productive at higher altitudes, more of it is needed to yield a given return. Because of this factor, upland holdings had to be larger to produce the same total harvest as lower-altitude estates. Furthermore, the investment of labor had to be greater, since the amount of labor required in agriculture is related to size of the area farmed: about the same amount of work is required to plow, manure, cultivate, and harvest a hectare of low-yield land as one of high yield. Peasant holdings in high valleys thus tended to be larger than those in the lowlands. However, the area a peasant household can work is limited, and larger holdings can compensate for lower productivity only up to a point. The net result was that the productivity of the upland holdings did not match that of the lowland estate, and the surplus available to the elite was accordingly less.

The ecology of mountain agriculture in high valleys did not favor the establishment of large, manorial estates worked by serf labor. Not only was the productivity of the land low, but the topography of the valleys is such that it is rarely possible to establish a compact mass of contiguous parcels of cultivated land. It would therefore have been difficult to provide the close operational supervision such estates require. Instead of the compact, nucleated community characteristic of the lower valleys, the new settlements in the high valleys were made up of scattered homesteads, providing each householder with the best possible access to his lands. Moreover, most of the newly settled valleys and plateaus did not contain within their confines sufficient land to support enough households to make up an adequate domain. The total rents to be derived from a single high valley were rarely sufficient to support its own nobleman. Therefore, many were simply included within the domain of a lord whose home continued to be a castle in the lowlands.

Low productivity, a dispersed settlement pattern, restricted size, and remote location all contributed to the erosion of feudal patterns of managerial supervision in favor of peasant management. The peasant proprietor came to make the crucial decisions on how production was to be carried out, and, in doing so, gained control over his harvest. The overlord could no longer make detailed estimates of the produce of each estate and of the precise consumption requirements of each household. Instead he was forced to make do with the collection of a fixed, hereditary rent from each estate, in return for which the proprietor gained rights to his estate in perpetuity, for himself and his heirs. How the peasant managed his resources was now up to him—however, he could still be removed from the estate for mismanagement. Although in lean years he might have to tighten his belt, or defer a portion of his rent payment against the hope of a better harvest the next year, in fat years the peasant could use his surplus to his own advantage.

The mountain peasant thus gained considerable freedom to regulate his own affairs. Moreover, this freedom was enhanced

"As former pastures and forests were converted to plowland and meadow, still more marginal land on the mountainside was pressed into use." High pastures, Gröden/Val Gardena.

further by the necessity of acting in consort with his fellow villagers to manage communal grazing lands and forest. The village council, once established, led to further political leverage since it could be put to uses other than those for which it was created. It interposed itself between the nobility and the individual households, becoming the vehicle for presenting community interests to the outside world.

Yet, it should not be thought that the mountain peasant was able to free himself immediately and completely from the exercise of domain by the lord. While the high valleys formed a frontier zone for settlement by peasants practicing mixed agriculture, these pioneers were not moving into an area outside of lordly domain. The high valleys did not constitute a frontier in the same sense as did, say, the American West in colonial times, an area beyond the confines of the existing state, free for the taking; they were already included within the political jurisdiction of the existing feudal order. Settlement there represented a change in the pattern of land use, rather than the opening up of virgin lands. It represented a change from seasonal habitation based upon the grazing of animals from lowland villages to continuous habitation supported by mixed agriculture. In moving to the high valleys, these peasant settlers needed support against the prior claims of lowland communities already using the land as summer pastures for their animals; even to this day mountain communities continue to share their pastures with lowland villages. Furthermore, as former pastures and forests were converted to plowland and meadow, still more marginal land on the mountainside was pressed into use to replace these usurped pastures and to provide lumber and firewood. On this high ground prior jurisdiction was often not clearly established, and as it was brought into use, neighboring villages often came into conflict over rights to this land. In this too, the new communities required the support of their sponsoring elites against the counter claims of their neighbors.

THE COMMERCIAL REVIVAL

The revival of transalpine trade over the Brenner and Reschen routes, discussed in Chapter II, coincided with the settlement of the high alpine valleys. For a while the Tyrol resumed the role of Rhaetia in Roman times, providing services necessary to keep goods moving, rather than serving as an independent source of trade goods. The transportation of this merchandise through valleys and over mountain passes depended on facilities for overnight halts, temporary storage houses, and a labor supply, including craftsmen. In response to these needs, towns grew up along the routes of trade—inns catered to merchants, and a variety of tradesmen participated in the transport of their goods. These towns have continued to flourish and have grown into the modern cities now found in the Tyrol. Later, roads were improved and new ones constructed, flood control programs were initiated along the routes of trade, and fairs were established—all signs of a political climate agreeable to the movement of merchants and their goods.

However, conditions favorable to commerce were not established all at once. Rather, they were worked out over a period of several hundred years, developing in direct proportion to the success of political consolidation. The growth of transalpine trade and political consolidation were inextricably interwoven: each increment in the development of commerce stimulated further political consolidation, which in turn resulted in still more trade.

The growth of towns in the Tyrol, as well as the trade moving through them, led to the development of new sources of wealth and power. It had not been possible to develop an effective centralized political organization as long as the peasantry alone supplied nearly all of the available wealth. The commercial sector of the economy, however, as small as it may have initially been, soon provided wealth not derived from agriculture, and that was free from the control of the landed nobility. The nobles who acted in concert with the interests of commerce were able to expand their hegemony over their land-oriented peers to become the new territorial elite. The most successful of these, the Counts of Tyrol, were unknown to historical records before the eleventh century. Yet, from a beginning as holders of a small domain near Meran, they eventually became the power that united the Tyrol into a single political unit. In the process, they successfully contained the power both of the landed gentry and of the ecclesiastical rulers of Brixen and Trento.

Understandably, the local nobles fought this process of centralization, but the territorial bureaucracy first established by the Counts of Tyrol and later amplified by their successors, the Habsburgs, effectively chipped away at their privileges and power. New judicial posts were created and assigned to local representatives of Bishop or Count. If these were sometimes held by members of the traditional nobility, they no longer held them as a hereditary due, but rather as a function delegated to them by the growing territorial state. Settled communities, both urban and rural, were granted charters that allowed them to regulate their own affairs and to bypass local nobles in their external political relations. The charters were managed by locally elected officials and councils, answerable ultimately to the territorial ruler, and no longer to the local, noble elite.

The territorial ruler and his officials, moreover, strongly supported the ongoing trend toward peasant-managed estates. The fixed rents owed to local lords declined in worth as an increasingly inflated economy reduced the real value of money. As their hold over peasant communities decreased, the nobles were in no position to increase peasant rents to compensate for inflation. Many nobles were, therefore, less reluctant than in the past to cede control of peasant land. The nobles who did not succeed in carving out a career in the imperial administrative apparatus, and who remained wholly dependent upon land, faced economic ruin and had no alternative but to take up the plow themselves. The impoverished local nobility was no match for a territorial government that could draw on the revenues from trade and mining and that increasingly filled its coffers from peasant taxes. As we shall see in Chapter V, local communities in the Upper Anaunia were granted communal charters by the end of the fourteenth century. Peasants in the area were increasingly successful in buying full rights to their estates in the succeeding centuries. This did not mean, of course, that they were freed from paying dues and taxes. Many of them continued to pay the tithe to the Church, and, increasingly, the representative of the central government demanded access to peasant surpluses of produce relinquished by the local nobles.

Not only were the peasants freed from feudal obligations, but the growth of towns along trade routes created a market for their produce. In response, the Tyrolese hinterland took its first steps in the direction of commercial farming. To be sure, most of the countryside clung to a subsistence economy—continuing to raise the full range of traditional crops and livestock. But the opportunity to sell portions of the wheat, rye, garden vegetables, pork, beef, and dairy products was now there, and surplus goods could be marketed in the towns. In the lowlands, however, villagers with ready access to the towns were going a step further in market involvement. A reevaluation of traditional production was taking place in order to identify products that would be most profitable. Vineyard and orchard crops, as well as cattle, proved to be the most promising commodities. Plowland was converted to orchard, and since one could make hay amidst the trees, the orchard served as meadow, too. Meanwhile, on less level land, vineyards expanded at the expense of field. Food and fiber no longer produced at home, but still required by the household, could be purchased from peasants in the uplands, where climate precluded the planting of orchard and vineyard. Yet, the trends toward specialization in the lowlands and a growing economic interdependence between lowland and upland estates, which were then beginning, would accelerate in modern times.

THE EARLY MODERN PERIOD:
SIXTEENTH–SEVENTEENTH CENTURIES

As we have seen above, the forging of a Tyrolese polity and economy was but one episode in a larger process of nation-building. Before the fourteenth century had come to an end, the Counts of Tyrol were succeeded by the Habsburgs, and the Tyrol became a crown land of the Austrian Empire. The initial value of the Tyrol to the Habsburgs lay in the strategic importance of the high alpine passes and in the approaches to these passes, through which the monarchy gained access to its western and southern provinces. At the same time, the output of Tyrolese silver and copper mines was expanding, and mining soon took on critical importance in the political economy of Austria. We have traced the social and political effects of this development in Chapter II. Its economic impact on the Tyrolese countryside was equally profound.

Mining had begun in the vicinity of Trento by the end of the twelfth century (Tremel 1969: 163), and over the next few hundred years the number of mines and their output steadily increased. At first, prospecting, operation of the mines, and smelting of the ore were carried out by local entrepreneurs, financed by local capital and regulated by the territorial elites. But these activities made sudden progress in the fifteenth century and continued expanding into the second half of the sixteenth century. As the scale of mining expanded, and as deeper deposits were tapped, control of mining operations was taken over by foreign capitalists, who operated on a grander scale and with more efficient techniques. Simultaneously, the regulation of mining and the fiscal benefits to be derived from it passed from the hands of local to national elites, as the mines became one of the most important resources of the Habsburgs and contributed significantly to their quest for empire.

Yet, by the early seventeenth century, ore deposits were giving out, and those that remained were ever more difficult to reach. Thus, production costs rose. Narrowing profit margins forced small operations out of business, leaving a few large firms to dominate the industry. Finally, rising costs bankrupted even the large firms, and many mines were closed for good. Others were taken over for a time by the crown, which continued to operate them as a means to prevent wholesale unemployment (Tremel 1969: 169). Thus, mining, once a signal asset to the monarchy, became in the end a severe liability. By 1630 even the state mines were closed down, and mining ceased to be a dominant factor in the Tyrolese economy.

Contrary to expectations, the mining episode in Tyrolese history did not usher in a period of development and progress, despite a marked acceleration in the scale of economic activity. The profits produced by the industry were siphoned off by foreign capitalists and by the agents of the monarchy. Expanded mining unleashed market forces which deranged the patterns of rural life, without—at the same time—offering economic alternatives to the rapidly expanding rural population.

Before expansion, mining was a peasant sideline. Most of the labor needed in the mines was provided by peasants who worked in the mines seasonally, or at intervals throughout the year. As the scale of mining operations increased, professional miners began to migrate into the Tyrol from Germany and even from Belgium, in response to the rising demand for a permanent labor force. Because these men were skilled miners, they had a competitive edge on the less skilled, local labor. The newcomers thus forced the part-time peasant labor force to surrender an important source of subsidiary income.

Other outlets that might have absorbed a growing population proved equally inadequate. Peasants continued to migrate to commercial centers such as Innsbruck, Bozen, and Trent. The growth rates of these towns were low, however, and a part of their growth was due to "natural" population increase. They were able, therefore, to accept only a small percentage of job-seeking immigrants. In addition, the upland frontier, which had served to ease population pressures in the past, was now closed. By the fourteenth century, communities had been established in all upland areas, and they could no longer receive migrants from the lowlands. For a time, new estates continued to be established by native sons, but by the sixteenth century, the high valleys were no longer able to absorb even their own population growth; thus, migrants from the uplands joined those from the lowlands. Population pressures became general throughout the Tyrol, and competition for the means to a livelihood ever more keen.

With the frontier closed and neither the mines nor commerce able to absorb excess population, the jobless returned to the countryside. Pressures to fragment holdings grew. These pressures were compounded by a vastly increased demand for agricultural products in the booming mining towns, and by the mines' need for goods such as wood, charcoal, and leather, and for services like hauling, smithing, carpentry, and masonry, which could be provided by peasants. Whereas a subsistence holding could not be divided without destroying its economic viability, division was feasible if one had a supplemental source of income. An expanded demand for rural products thus resulted in the proliferation of dwarf holdings.

The problem of fragmentation was further aggravated as the foreign miners established themselves in the Tyrol. Not far removed from their own rural origins, they saw in the possession of land— rather than in wages earned from labor—the ultimate source of material security. Efforts of this proletariat to establish itself on the land were intense and contributed further to the trend toward fragmentation of estates and to the general deterioration of the rural economy.

TYROL AND TRENTINO CONTRASTED

The growth of long-distance trade, accompanied by the founding of towns, the penetration of a national market economy, and the advent of mining industry, all combined with the closing of the upland frontier to shape the Tyrolese rural economy. The results of these pressures were first, a shift from production aimed only at meeting household requirements to a growing interest in market opportunities; second, a concomitant increase in economic interdependence within the countryside, as some estates, notably in the lowlands, narrowed their range of production to only a few marketable items; and third, demographic problems in the countryside, which were expressed in the division of many estates into small holdings.

These conditions were equally present in both German and Romance-speaking areas of the Tyrol, but the responses in the two areas differed and continue to differ today. The German Tyrolese, with their concept of the self-contained, perpetually indivisible estate, resisted fragmentation of peasant holdings. Successful adaptation to the market required a holding large enough to support the household from the sale of its produce, or from a combination of sales and home consumption. Whether oriented to subsistence or to the market, the Germans continued to view division as a threat to economic survival. This definition of the situation and the strategy based on it found support among the regional elite. The aim of the elite was to achieve a high level of agricultural production *per household,* which would be available to provision other sectors of the economy, and which would also provide the highest tax per unit of land. Since the members of the regional elite and the estate managers agreed on the desirability of maintaining indivisible estates as a precondition to their survival, both practice and law resisted division of property.

Yet this goal could be attained only if population growth were met by expanding economic alternatives; as we have seen, such alternatives *disappeared* at this time. Thrown back on the countryside for support, a growing population pressed its own claims against prevailing law and practice. Whereas the image of an estate large enough to meet all the needs of a household might serve as an ideal for a successful peasant owner or a member of the elite, a countryman with no prospects in either town or country would be inclined to fight for any available resource—a fragment of land— that could provide some income. Under such pressure, division of properties became more frequent, producing many estates too small to support a household. Managers of reduced holdings found themselves, soon after acquisition, pressed to supplement their incomes.

In the Trentino, the dominant strategy for securing adequate economic units emphasized the viability of households, rather than the viability of estates. The household head was concerned less with the amount of land in his possession than with the total range of his resources, including marketable skills. The overlord's interest ran parallel: as long as the household could pay its taxes, regardless of the source of income, he was satisfied. In this rationale, more households meant more taxable units. As a consequence, policy makers in the Trentino did not view the division of estates as an economic and political tragedy.

The consequences of these two different policies proved roughly similar, as long as production remained oriented toward subsistence. Under such conditions, household viability and estate viability meant the same thing. As money became the medium of exchange and market activity increased, however, a household could survive by combining farming with a saleable trade or craft; with sufficient expansion it was even possible to survive simply by trade. Of course, not all estates were divided—many survived by farming alone. But resistance to division was lower in the Trentino than in the German-speaking Tyrol; under similar economic and ecological conditions more estates were divided in Romance, than in German, communities.

If two households—one in a German community, the other in a Romance-speaking village—each combined farming with, say, carpentry and did so in equal measure, an outside observer might easily be tempted to regard them as identical economic units. Tyrolese and Trentini, however, would immediately draw a distinction between them, a distinction that reflects different patterns in the division of labor within the villages and has bearing on the direction of economic change. The German would regard himself, and would be regarded by others, as a *Bauer* (peasant) who, because of insufficient land, was forced to practice a sideline (*Nebenbeschäftigung*). The Romance household head, on the other hand, would be likely to insist that he practiced a profession (*professione*) and only engaged in farming because there were too few clients. The German peasant prefers to be a "jack of all trades" and the master of an independent household, whereas his Romance counterpart emphasizes a single skill and relies on his neighbors to provide complementary services, playing down his peasant (*contadino*) status in the process. Thus, the degree of interdependence, then as now, has tended to be greater among households in Romance villages than among the German speakers.

While this divergence was not marked, as long as households continued to produce mainly for their own needs, rather than for market, growing involvement with commercial transactions quickly amplified the difference. Given the opportunity, the German Tyrolese is inclined to increase his holdings and give up his sideline. The Trentine cultivator, on the other hand, is more inclined to abandon the land entirely, as soon as he can support himself through a profession. Thus, the contrast in economic organization of the countryside between the German Tyrol and the Trentino has widened as the market has penetrated ever more deeply into rural areas. It remains least pronounced in upland regions such as the Upper Anaunia, where market production has remained relatively marginal into recent times. We shall explore some of the local manifestations of these phenomena in Chapters VI–VIII.

THE ECONOMIC MARGIN OF THE STATE

The sixteenth and early seventeenth centuries saw an incipient industrial capitalism in the Tyrol collapse with the end of the mining boom. While there were unique reasons for the termination of capitalist growth in the Tyrol, the entire south-central German-speaking area witnessed a general decline of early capitalism, as the center of gravity of capitalist development shifted to north-western Europe. Industrial growth eventually revived in Austria, but now in the eastern provinces. The Tyrol was never again to gain a position of national economic prominence, neither under the Habsburg, nor under Italian, rule. To this day, the contribution of the Tyrol to the gross national product is minimal.

Nevertheless, the Tyrol has not failed to become firmly linked to the national economy. Although mining collapsed and the relative importance of transalpine trade to the monarchy diminished, trade continued to grow in volume, and the integration of the Tyrol into the Austrian Empire and its relations with greater Europe came to control the direction of Tyrolese economic development. Two factors stand out in the external economic relations of the Tyrol in the period 1600–1800. First, there was the growing responsiveness of Tyrolese rural production to the pressures of the market, both in the sale of agricultural produce and in the development of rural crafts. Tyrolese products, especially wine and fruit, were marketed increasingly outside the Tyrol, and external producers began to compete with Tyrolese produce in Tyrolese markets. Heightened involvement in the European market system subjected Tyrolese commodities to competition from the outside, a problem that proved especially acute in the production of grain crops.

The second factor was the migration of Tyrolese to other parts of Europe, and, late in the nineteenth and twentieth centuries, to the Americas. Without internal industrial growth, the Tyrol could not gainfully employ all its sons and daughters, and thus became dependent on the rest of Europe to absorb its overflow population. Begun in the sixteenth century, this migration grew in volume through the seventeenth and eighteenth centuries, reaching its peak between 1880 and the 1920s. Emigration continues as an element in the Tyrolese economic equation even today.

Many left the Tyrol as permanent migrants, taking up new lives in other parts of Europe, especially within the Austrian Empire; others left the Tyrol seasonally, returning after a few months of wandering. This was especially true of the victims of previous centuries of land division who held too little land to support their families. A common practice was to carry with them objects manufactured on the holding during the winter months to sell, here and there, in other lands, perhaps buying and selling additional objects along the way. The traveling merchants of Tyrolese origin came to be a common feature in the marketplaces of Europe during the seventeenth and eighteenth centuries. In addition, many of the migrants carried with them not goods but skills—especially in the building trades, and they would travel to places where building projects offered seasonal work. Still other Tyrolese, possessing only peasant skills, traveled where they could work as farmhands, as herdsmen or, on occasion, as lumberjacks.

Farmhands were especially in demand in the farmlands of southern Germany, especially in Swabia. The migrants to these lands included not only men, but also women and even children. Small holdings with large families could survive an inadequate harvest by sending the smaller children elsewhere to be fed. Leaving their homes on foot in the spring, they would arrive after some weeks in Swabia where they would labor through the summer months caring for livestock, sometimes helping in the fields, or if fortunate, serving in the house of a prosperous peasant. In the fall, they would return to the Tyrol on foot, bringing a few coins in their pocket. Their meager earnings contributed little to the household economy, but their absence during the summer season meant that more of the family harvest could be stored for use in the winter months (Wopfner 1951–1960, Vol. II: 384–403).

Combined with patterns of late marriage for landowners and celibacy for the bulk of the landless, which served to inhibit population growth, migration relieved the population pressures that had led to the deterioration of farming during the mining era. This was not accomplished by a complete removal of excess population from the rural scene, although some migrants did manage to find permanent careers in distant lands. Many more spent the major part of their lives as nomads. Small holders migrated or sent their sons and daughters away on a seasonal basis in order to make up the difference between what the land could provide and what the family required to live. Others who controlled no land might seek sustenance and shelter for a part of the year with a landowning brother, especially if that brother needed additional labor from time to time— but much of their time was spent on the road. Migration has been as much a part of life in the Upper Anaunia as elsewhere in Tyrol, so we will discuss it and its relationship to local social and economic organization in some detail below.

CHANGING ECONOMIC RELATIONS

As market relations penetrated the countryside, estate managers were increasingly relieved of the necessity to provide for all their needs directly from their own land. Simultaneously, land and labor became commodities, and pressures mounted to reevaluate operational strategies so as to maximize profits. We have already shown how a mixed agriculture, directed toward household requirements, gave way to ever more specialized production designed to satisfy market demand. Such commercial pressures mounted slowly and only sporadically until the nineteenth century. Since then they have grown exponentially, and within the past twenty years have become ubiquitous and unremitting. Still, market expansion has not proceeded at a uniform rate, nor has it affected all parts of the region in the same way.

The overall tendencies were toward a shift from crops grown to satisfy household demands toward crops grown for market sale, and an accompanying "rationalization" of production. When self-sufficiency in production was the chief goal of land use, maximum productivity of land was not the primary consideration. It was then of no concern if land that could provide excellent stands of wheat was used to grow fodder, as long as the wheat harvest met the needs of the local population. Similarly, if land was not good for growing wheat, but well suited for livestock, a minimum amount of land would be devoted to raising wheat for local needs. But once production for market became of greater concern than household self-sufficiency, land-use strategy dictated that each producer direct his efforts toward finding what he could produce most profitably. Since the range of food and fiber required by the growing towns was much the same as that of the countryside, it was necessary only to decide which of the crops already being produced to grow. While a few minor crops might be dropped from production altogether, it was unusual for a community to abandon any of the principal crops or types of livestock completely. Rather, mixed agriculture continued, but with shifts in emphasis on the crop or animal that seemed to show greatest promise, given local conditions, for market returns (Figure 4 illustrates schematically the resultant patterns of land use). Not all communities, however, came equally under pressure from the market. Market involvement was constrained by prior commitment to subsistence production, by variations in ecological setting, including the degree of isolation, and by prevailing political conditions. Generally speaking, it has been the holdings located at lower altitudes, in the main river valleys, that have been most affected by the market and continue to balance market and subsistence strategies. Along this altitudinal continuum we can distinguish three patterns of market involvement.

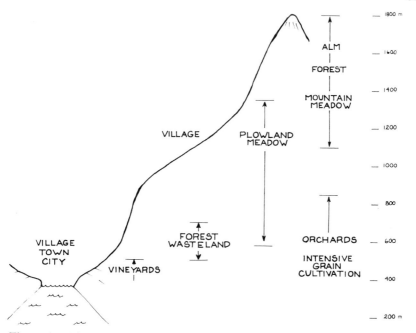

Figure 4. Schematic diagram of land use: Trentino–Tiroler Etschland region.

In the lowlands, heavy capitalization of agriculture has been invited both by substantial productivity and by improved transportation networks. These have made distant markets available and have brought in goods produced in other parts of the Alps and outside the Alps. Production on each holding is concentrated on a single product to the virtual exclusion of all others; output is sufficient both to fill local demand and to provide each region with an export commodity. As a consequence, production is oriented toward national and even international markets. South of the Brenner, in the warmer corridors of the Etsch River and some of its main tributaries, vineyards and orchards cover the land, and the major market commodities are wine, fresh fruit, and preserves (Leidlmair 1958, 1965a, b). In all areas, cooperative, as well as privately owned, plants process and market local specialities and provide a good living for local entrepreneurs.

Over the past hundred years, much land has been brought under cultivation in these lowlands in response to market pressure. Swamps and marshlands have been drained, and expensive flood control programs established. Land made available through these programs has proved to be as productive as any other lands in the

"In the warmer corridors of the Etsch River and some of its main tribu-taries, vineyards and orchards cover the land." Vineyards near St. Paul Eppan/ S. Paolo Appiano, in the valley of the Etsch, west of Bozen. In the background, the ridge of the Mendel.

Alps. Although some of this land has simply been incorporated into existing holdings, entirely new estates have also been organized. The newly established estates are wholly oriented toward produc-tion of cash crops for market, and have never participated in the ancient pattern of mixed agriculture. Indeed, with their land parcels confined to river-bottom land, the estates lack access to the com-bination of land types that is the hallmark of mountain agriculture.

In estates of more ancient derivation, old patterns of land use were greatly modified. In vineyard and orchard areas, plowland and meadow were totally converted, the harvest of mountain meadows ceased, and rights to graze livestock on high pastures were aban-doned since livestock are no longer kept.

The second pattern of relations is found in the high lateral valleys of the Tyrol, and in the upper reaches of some of the main valleys. Altitude and climatic conditions combine to preclude the growth of vines or fruit trees, and to inhibit intensive grain or hay cultivation. Here mixed mountain agriculture remains, but strategies of production are now oriented toward supplying regional markets. The combination of plowland and meadow has shifted away from the balance demanded by subsistence production to reflect the particular market produce being grown. In general there is greater emphasis on field crops at lower altitudes and on livestock-raising at higher altitudes. Thus, the percentage of home-field parcels under the plow decreases as one moves up the mountainside. This reverses the old pattern of subsistence farming. When each family had to grow all its own field crops, households at higher altitudes required a higher percentage of land under the plow than did those at lower altitudes because of lower yields per hectare in the uplands.

In the cattle-raising areas, plowland not devoted to feed crops has shrunk to the size of kitchen gardens, as meadowland is expanded to maximize hay harvests. In many communities, both the harvest of more marginal meadowland high on the mountainside and the use of summer high pastures have declined, or have even been abandoned, in the interests of improved quality of livestock and efficient use of labor. At the same time, competition for high-quality meadowland increases as all seek to expand their scale of operations.

Households in this category continue to provide for many of their own needs; they rely, however, in large measure on the sale of produce to lowland markets. The cities and lowland farmers will consume all the dairy products and field crops they can provide. Still, these estates are discouraged from full commitment to the market by lower levels of productivity and by their heritage of subsistence agriculture. However, these households can make a better living through a combination of subsistence and market farming than through raising cash crops alone.

The third pattern of adaptation differs from the middle range not in what is grown, but in how this produce is used. On estates situated high on a mountainside or at the upper extreme of a minor valley, production remains subsistence oriented. On holdings near the ecological limit of cultivation, where there is a short, cool, growing season, both the quality and quantity of produce is even lower than on the holdings located at middle-range altitudes. These estates have little to offer the market, and the market has little to offer them. All-weather roads, built to connect middle-range communities to the lowlands, have only occasionally been extended to these communities of scattered estates at highest altitude. Thus, they remain marginal to the market: cash requirements continue to be minimal and are met by the sale of produce during infrequent trips to market, or by temporary periods of wage labor in the lowlands.

An interesting aspect of this marginal market relationship is that although production remains subsistence oriented, land nevertheless has become a commodity. Demand is based not on the market value of the output of the estate, which is minimal, but on the ability of the estate to support a family. Disinherited sons of the land, who must work as wage laborers but want holdings of their own, compete for estates as they become available. Thus, a market for subsistence estates exists, driving their price well above a "true" market value.

THE FASCIST INTERLUDE

The economic organization of the countryside, that we have set forth lasted well into the twentieth century, and has been seriously modified only within the past decade. The traumatic episodes of European history in this century have, in the main, encouraged economic conservatism in the Tyrol: the land can continue to support its owner, regardless of the rise and fall of external institutions. Manpower was siphoned off the land and assembled to do battle in the Great War, so that the labor of those who remained had to be increased. Many returned from the war, but the numbers from the Tyrol who did not return were large enough to be reflected in demographic charts: the generation that had the misfortune of growing to maturity between 1914 and 1919 was severely decimated. On the other hand, there were few battles within the Tyrol itself, so that production was not disrupted.

The war's end found the Tyrol divided, with the Trentino and the South Tyrol transferred to Italian sovereignty. This had little effect on day-to-day farming operations, but it completely upset the market system that had absorbed goods produced for sale. With the Austrian Empire dismantled and an international boundary interposed between them and their traditional customers, farmers south of the Brenner had to find new markets for their wine and fruit. In time, the necessary adjustments were made, but soon after there came the economic hardships of the twenties and thirties.

The depression was keenly felt throughout the hills and valleys of the Tyrol. Once again, the land was called upon to support those who could not find work, as jobs grew scarce in Europe, and many who would have migrated had to remain at home. Moreover, all estates were adversely affected by cash incomes, which sank in relation to need. Small holders who relied on a trade, craft, or income from outside labor were especially hard hit. Debts were pressing, and not all estate managers avoided bankruptcy; many estates changed hands, and others were broken up and sold piece by piece. Overutilization of forests by peasants in need of cash left many a mountainside devoid of commercially valuable timber. Throughout the Tyrol, the long-term trend toward increased market involvement was temporarily arrested: peasants fell back on subsistence production as marketing became unprofitable.

World War II also took its toll on the Tyrol, although there were no land battles within its borders, nor any bombing raids until late in the war. Again, men were pressed into military service, and countrymen were faced with forced deliveries, rationing, and other economic hardships. These hardships, however, left the major outlines of rural life untouched, and a depressed economy near the end of the war and in the years following again showed the Tyrolese the prudence of subsistence farming.

Of more lasting significance were economic changes introduced by the Fascists during their twenty-one year control of Italy. From 1922 to 1943 they undertook to Italianize the Tyrol and to suppress Germanic nationalism, language, and culture. While political and cultural measures were emphasized most, the economic aspects of Fascist policy proved to have the most lasting effect. Initial Italian economic plans for the Tyrol centered on the development of hydroelectric potential: the first hydroelectric plant was constructed in 1924 at Marling, near Meran, and the energy potential of Tyrolese streams has since been increasingly exploited. Within two years, industries began to develop: the first of these was a chemical plant built in 1926 by the Italian firm of Montecatini. Early efforts at industrial development were initiated by private industry, but the scale of development expanded in the mid-1930s as a result of a government decision to establish an industrial zone in the midst of the German-speaking region.

This decision linked the Fascist effort to Italianize the South Tyrol with a more ambitious plan to make Italy economically self-sufficient. Throughout the country, industrial expansion was under way, supervised by state planning agencies and financed mainly through state-controlled banks. It was hoped that an industrial zone established in a Tyrolese city and controlled by Italians would not only further Italian national goals, but also attract Italian migrants to the area. These migrants would lend authenticity to the picture of an Italian South Tyrol that the Fascists hoped to paint, and they could be useful as an additional force in assimilating the indigenous population.

One of the first steps was an Italian takeover of the banks in the region, including the *Bauernsparkasse*, or Peasant's Savings Bank, which had been the principal institution for the support of agriculture and small business development, and which was the region's main savings facility. With this and other banks in state hands, Italians were in control of virtually all capital in the province. Dominating both capital and economic planning, the Fascists were able to successfully control the direction of industrial growth; German speakers were effectively excluded from any participation. Managers and personnel were recruited in the center and south of the country; government propaganda extolling the opportunities to be found in Tyrol was combined with offers of economic assistance in order to induce Italian speakers to immigrate.

At the same time, the German-speaking population found it virtually impossible to raise capital for new enterprises and difficult to improve or maintain those already in operation. Moreover, the German-speaking population held no sway in an emerging industrial sector dependent on Italy proper both for supplies and markets. It was powerless to influence industrial growth in any way.

The main industrial center was established in a zone outside of the city of Bozen; this has remained the principal manufacturing area in the South Tyrol. However, capital and tax incentives were offered Italians wishing to establish businesses or factories elsewhere in the South Tyrol as well, and in response Italians intruded into the economic life of other cities, especially Meran and Brixen. As a result a new urban complex developed in the South Tyrol, fed by a continuing stream of Italian-speaking migrants from the south. After World War II, official policies discriminating against German participation in the urban complex were dropped, although government subsidies to industry continued to be necessary for a time. By then, however, the growth of the urban complex had a dynamic of its own. German speakers were reluctant to participate in this Italian-dominated economic sphere, whereas workers from all parts of Italy have continued to arrive and settle in Tyrolese cities.

In response to the attention given them by the Fascists, the growth of cities in the South Tyrol has been fed by an Italian, rather than a local, German, hinterland. In 1921 Bozen was a rural town of barely 25,000. By the mid-1960s it had passed the 100,000 mark. Before World War I, the number of Italians in Bozen had been about 1300, well under 10% of the total. In contrast, the 1961 census showed nearly 70,000 Italian speakers out of a total population of some 90,000, or nearly 80% of the total. Since the number of German speakers in the city has remained virtually constant since 1918, the total urban growth in Bozen can be attributed to the influx of Italians. Although the percentage of Italians in Meran and Brixen is lower than in Bozen, the pattern is the same. Two-thirds of the population in the South Tyrol's three largest cities were Italian as of 1961.

In marked contrast to Fascist success in controlling the growth of Tyrolese urban centers was the failure to settle Italians in the countryside. An organization was established in the 1930s to buy up farmland from Germans to sell to Italians, and financial assistance was extended to prospective rural settlers as well. Establishing a worker in an industrial job is one thing—establishing a farmer in a peasant village quite another. Farming techniques in the mountains were unfamiliar. Furthermore, cooperative relations with neighbors were crucial, but German Tyrolese were disinclined to assist *contadini* who accepted a role of aiding in the Italianization of the countryside. Ostracism and failure led many to remigrate. Ironically the few who succeeded in striking local roots did so largely because they learned to speak German and accepted their neighbors' modes of behavior and viewpoints. In so doing, they gained acceptance, but they did so by violating the goals set by the state.

Even the *Option* failed, and with it the plan to empty the land of nationalistic Tyrolese, leaving behind only a small residue of German speakers who would soon be assimilated by immigrating Italians. Few landholding Tyrolese actually migrated, however, and those who did frequently maneuvered to sign over their estate to a close relative, thereby remaining tied to the land. At the collapse of the Fascist government in 1943, the South Tyrolese countryside was as staunchly German speaking as ever.

Yet, after Germany's surrender, the South Tyrolese found themselves once again included within the Italian state. Whereas the countryside had remained German, the cities soon showed the effects of continuing Italian immigration. The "Old Town" in the center remained Habsburgian in character, but the "new quarter" became solidly Italian in appearance, and ongoing control of the urban economy remains in Italian hands. This contrast is not lost on the inhabitants of the South Tyrol. The bitterness between German- and Italian-speaking populations has been fed by economic policies begun by the Fascists and continued through inertia. Each group draws invidious distinctions between itself and the other. The German–Italian distinction now also carries a parallel rural–urban distinction. For the South Tyrolese, to be urbanized is to be Italianized; hence, he prefers to remain in the countryside. This has given a tremendous impetus to rural values among the German speakers: even small landholders prefer to protect their stake in the land although it may mean a living standard obviously lower than that of a wage earner. The landless seek trades and professions that can be practiced in the countryside, especially construction work and tourism-related activities; wage earners of rural origin seek to gain a toehold in a farming community—to eventually acquire a viable estate. If a Tyrolese must seek work in the city, he prefers to go to Switzerland or Germany, rather than to remain in the South Tyrol or to move to an Italian industrial center.

THE POSTWAR BOOM

During the 1960s the economy of both the South Tyrol and the Trentino continued to be dominated by farming, but industry and commerce played a greater role than ever before. In common with the rest of northern Italy and Europe at large, the region has experienced unprecedented economic expansion and prosperity. Opportunities to market farm produce have grown, and commercial farming, whether based on orchards, vineyards, or cattle-raising, can provide a good standard of living. Even in remote areas such as the Upper Anaunia, estate managers have responded to these opportunities. Everywhere true subsistence production is disappearing and, should the current pattern continue, will soon be but a memory.

Moreover, for the first time in the history of the Tyrol, numerous economic alternatives are available to those with little or no land. In the Trentino expanding cities and towns are able to absorb many of those who choose, or are forced by circumstance, to leave the countryside. Others find employment readily available in industrial centers elsewhere in northern Italy, Switzerland, or Germany. Furthermore, the expansion of income in the countryside has created many new niches for those who prefer to avoid the urban centers. Trucking, processing rural produce, the construction industry, and many other trades and crafts are real alternatives, and entry into these professions is now facilitated for the young by training available in a number of trade schools that have been established in recent years.

Above all, the tourist trade has grown in response to the leisure demands of the prosperous European middle and working classes. Always a favorite health resort area and playground for the well-heeled upper classes, the South Tyrol has witnessed the influx of the less wealthy experienced by most of Europe's resort areas in recent decades. Both summer and winter facilities have been expanded, and, while many of these are located in cities, many more are to be found scattered throughout the countryside—and communities vie with one another to build future resorts. In a few special cases, whole villages have turned to catering to tourists, and farming has been virtually abandoned. More frequently, a few local entrepreneurs derive their living in whole or in part from tourists, whereas others continue to farm. In any case, the expansion of tourism provides employment opportunities for many villagers, either in their natal village or in one of the large tourist centers.

During the early, leisurely growth of commerce in the lowlands, with labor in oversupply relative to the needs of the nonagricultural sector of the economy, there was always available to estate managers a labor pool made up of landless villagers and individuals who held dwarf estates. Peasants with medium and larger estates could count on all the labor they required in return for a minimal wage. As the nonagricultural sector of the economy expanded and its requirements for labor grew, however, wages climbed, and labor was drawn away from agriculture. This was especially true during the 1960s. Those without land left the village or, especially in this day of the motorbike and Fiat 500, become commuting workers (German: *Pendler,* Italian: *pendolare*) living in the village but working in town. These landless workers are joined by the small landholders, who can now better supplement farm income from urban, rather than agriculture, employment.

As landless villagers become independent of estate managers, the latter become increasingly dependent on the domestic force. Yet even the manager's children, as they approach maturity, tend to prefer the life of an independent wage earner to that of dependent farmhand on their father's estate. Many leave home just as they reach the age when they could make a substantial contribution to the estate's operation. This reduction of the labor pool presses on the manager's ability to operate his estate effectively, and makes the efficient utilization of available labor critical. In addition, the growing prosperity of wage earners provides the estate manager with a yardstick against which to measure his own standard of living. In particular, during the last generation, the wage of the city-based laborer has grown to rival and even surpass the income from average land holdings. Responding to market pressures, to invidious comparisons with the wage earner's income, and to his shrinking labor force, the estate manager redirects his use of land and labor to maximize his cash income.

The obvious solution to the problem of a shrinking labor supply is to mechanize estate operation and to introduce scientific farming techniques, which will raise productivity per unit or labor invested. The introduction of an elaborate technology requires capital: money will have to be invested and a cash income realized sufficient to bear the cost of equipment and to provide a living for the manager and his family. This move, once taken, is practically irreversible and commits the estate to a market economy.

As the mountaineers become increasingly committed to a market, they must reconsider the entire pattern of mountain agriculture. The range of produce will continue to shrink as managers devote ever more land and time to the crop or animal that will bring him the highest profits. Plowland expands at the expense of meadow when field crops are the specialty, and meadow grows at the expense of plowland when livestock-raising is emphasized. Moreover, the entire complex of mountain agriculture becomes less attractive when production is geared to cash crops. Mountain meadows are taken out of production as labor becomes more valuable than the fodder to be harvested. Yields from these meadows may be high, but machines generally cannot be used on mountainsides because of excessive slope and rock outcroppings, or because they are only partially cleared of timber.

The traditional pattern of cattle-raising in the Alps declines as the sale of dairy products and meat becomes more important. With the manager's interest now either in maximum milk output or high-grade beef, old patterns, which were designed simply to ensure herd survival, become untenable. Feeding must be carefully controlled and the milk cattle kept at hand to facilitate milking and milk collection. In many communities fully committed to dairy farming, patterns of transhumance have been abandoned and with them the traditional occupations of dairy maid and herdsman. Accompanying this trend has been the upgrading of village meadows through the use of high-yield fodder crops and commercial fertilizer, the establishment of village, or valley, marketing and processing cooperatives, and the rise of a breed of estate managers eager to acquire additional meadows in order to expand the scale of their operations.

CHAPTER V

History of an Upland Valley

We began this book with an attempt at characterizing the forces that first led to the unity of the Tyrol, and later subjected that unity to the strong, centrifugal pull of nationalist attraction. In Chapter IV, we discussed the ways in which these forces molded the economic fate of the South Tyrolese and Trentine peasantry. We can now see how the development and distribution of these forces affected the Anaunia, the upland valley within which St. Felix and Tret are located. The precise specification of events in this peripheral area awaits the inquiry of the professional historian, but documents available to us reveal something of the ways in which human action in the Noce River valley responded to the rhythm of economic and political development in the Tyrol as a whole.

THE ROMAN FRONTIER

The Anaunia was already a frontier province when the Romans drew a line across it in 15 B.C. to separate their long-established

Italic possession of Venetia et Histria from the newly conquered province of Rhaetia to the north. The twin valleys of Non and Sole were incorporated into the X Regio Italica, pivoted on Trento; and in A.D. 46 the Emperor Claudius granted the rights of Roman citizenship to Anauni, Siduni, and Tuliassi. The Anauni are obviously the inhabitants of the Val di Non; the Siduni have been identified with the natives of the Val di Sole; the identity of the Tuliassi is uncertain. The Clesian tablet in which the grant was recorded appears to have ended the threat of confiscation of lands and property by the victorious Romans. It states that the populations involved have long been in possession of their territories; that the status of their Roman citizenship has been uncertain, since some of them had not even been annexed to the region; that they could not be separated from Tridentinum without causing difficulties to the city of Trento; that many of them had in fact participated honorably in the military and legal life of the province. This tablet, therefore, sounds a note that was to recur throughout the history of the valley: part of a larger administrative region, with ties to Rome and to Latinity, it still remained marginal to the concerns of empire. The same is also true of later events that affected the course of empire in the Alps. The population of the valleys was still strongly pagan by the end of the fourth century A.D.; in 397, three Cappadocian missionaries, Sisinius, Martirius, and Alexandros, were martyred at Sanzeno, the site of a temple of Saturn. Yet, by Longobard times, in the seventh century, there appear to have already existed the beginnings of what later came to be the nearby pilgrimage center of S. Romedio. The area also remained somewhat marginal to the incursions of the Germanic tribes. Ostrogoth hegemony in the sixth century A.D. certainly reached as far as Trento; Frankish armies destroyed nearby Tisens, Sirmian, and Eppan on their march toward Verona. Only the Longobards appear to have left some traces of strongholds in the valley itself—at Romeno, Castelfondo, and Tres near Vervó. Carolingian forces may have passed through the Val di Sole, but they left the Anaunia lying to one side—undisturbed.

BISHOP AND COUNT

We know little about the process of consolidation that permitted the Bishops of Trent to build up an ecclesiastical domain comparable to the political unit ruled by the Counts of Tyrol. By the year A.D. 1000, the bishopric was able to sever its ties with the Margrave of Verona and put itself into direct relationship with the German emperors, a step welcomed by them as a guarantee of free access to the route into the Po River valley to the south. The Anaunia was formally divided into five administrative districts (*gastaldie*), with their respective seats at Cles, Romeno, Malé, Ossana, and Livo. These *gastaldie*, in turn, divided into *scarie* and *decanie*. In the thirteenth century, administration was centralized in the office of a *vicedominus;* somewhat later, juridical matters were delegated to a special officer, the assessor.

Yet this sketchy administrative structure was not always able to cope with the social forces it was designed to administer. Episcopal administration was often at loggerheads with nobles striving for preeminence, and with rural communes seeking to secure a measure of autonomy from both the Bishops of Trent and the nobility. Some noble aspirations were granted impetus by the administrative structure itself: as ecclesiastical ruler, the bishop frequently had to yield real power to secular lords, as protectors of episcopal interests. The first episcopal legates of this type in the Anaunia, called *Vögte* in German (singular: *Vogt*) and *avvocati* in Italian (singular: *avvocato*), were the Counts of Flavon, whose noble ancestry predated the formation of the Trentine episcopate. When their fortunes began to decline in the twelfth century, they yielded power to the Counts of Eppan, whose center of dominion lay in the vicinity of Lana and the Valley of Ulten, north of the Gampen Pass. Powerful and persistent colonizers as they were, in the course of the thirteenth century, they had to yield, in turn, to their fierce adversaries, the Counts of Tyrol, to whom they surrendered, in the Anaunia, Cloz in 1231 and Vasio in 1248. In 1250, the Counts of Tyrol became the formal legates of the Bishops of Trent and extended their sway over additional strong points in the Anaunia: they occupied Castelfondo in 1265, Sta. Lucia in 1271, and Arz in 1281. Their legal overright, however, could not contain the ascendancy and power of numerous, smaller, noble families in the Anaunia, who fortified their own castles and often fought bloody feuds, sometimes fanned by the conflicts between the Counts of Tyrol and the Bishop, sometimes by conflicts of both Count and Bishop with external powers. Some of these families were of Germanic origin, like the Thun; others, like the Cazuffi of Tuenno or the Concini of Malgolo, were Romance. Conflicts between the noble lineages came to a climax in the fourteenth century, coinciding with the period during which the Counts of Tyrol welded the entire Tyrol into a viable, unified, political unit for the first time. A great feud, 1335–37, saw the Thun, St. Ipoliti, and Tuenno ranged against the Arz, Cagnò, Caldes, Spaur, Nanno, Coredo, and Valer; in 1371, battles were waged between Thun and Thun, and between Arz and Arz. Both were ultimately suppressed by the military power of the Counts of Tyrol; but the fury of feud violence may well have been responsible for the old adage that "if ten devils start a fight with a Nónes, the Nónes will be victorious."

COMMUNES

Other political actors in the Anaunia were the rural communes, supported by Bishop and Count against the nobles in their assertion of the right to local self-government. Each commune had its charter of rights (*carta di regola,* German: *riegl*), and elected its own council. Fondo, the center of the Upper Anaunia, for example, has had an assembly of households heads since the end of the thirteenth century, and a charter of rights since the mid-fourteenth century. Written down in 1398, it was revised periodically thereafter with an eye to keeping out foreigners, regulating access to communal woods and pastures, and preventing the sale of fixed property in the commune to nobles. Through the use of this instrument, the commune successfully contested the pressures of the nobility during the fifteenth and sixteenth centuries, and defeated the claims of the Count of Thun, whose castle was at nearby Castelfondo. It allowed him entry into the town council as a citizen with one vote, equal to any other voter. At the same time, the commune as a whole owed Castelfondo payment of the tithe, some harvesting services, and obligations to deliver wood to the castle, to transport wine from the vineyards of Revó and Dambel, and to deliver specified amounts of meat, milk, and eggs (Inama 1905: 300).

The commune governed itself through three *regolani*, elected annually, and apparently representing the three sections of Fondo, San Martino, Forapónt, and Moích. The communes, with their *regolani minori*, in turn, formed a *communitas*, roughly comparable to the parish; like the communes, the *communitas* held annual assemblies and elected a representative, or *regolano maggiore*, to administer domestic affairs, and a *sindaco* to manage external affairs. Since the Bishop often wanted to place men of his own in the communal and parish offices, the representatives of local units and episcopal administrators were frequently at loggerheads. Communal and episcopal institutions also combined—and conflicted—in the economic sphere, governed by the so-called *magistrato*, with its seat at Revó. This committee consisted of the *vicedominus* and the assessor, as representatives of the Bishop, and three *sindaci generali*, elected by the communes. The *sindaci* were charged by the communes with the defense of valley rights (*privilegi delle Valli*) against encroachment by episcopal administrators.

GERMAN SETTLEMENT

The consolidation of the Bishopric of Trent and of the domains of the Counts of Tyrol also produced the first incursion of German speakers into the upper reaches of the Anaunia. Romance place names in the valleys of Pescara and Novella make it likely that this mountain rim of the drainage basin of the Noce was already occupied by Romance speakers before German settlement. Some place names suggest the use of local resources for pasture and firewood: Martscheinen (from Romance *malga casina*, transhumant pastoral hut) in the Valley of the Pescara; Senale (from *casinale*, site of transhumant huts); Pizuvit (from *picetum*, fir forest, until 1619 in use as a term for the neighborhood of Obere in Unser Frau); Malgasott (perhaps from *malga t'sot*, lower mountain pasture); Caseid (from *casetta*, hut; now St. Felix) in the valley of Novella.

It seems unlikely that any permanent settlement existed in the area before A.D. 1000. By 1184, a hospital—referred to as *Eccl. s. Marie de Senali*—was built at Unser Frau/Senale; the papal bull of that year freed it from jurisdiction of the parish of Sarnonico and placed it directly under the Bishop of Trent. (The hospital and associated monastery stood at the site of the present church of Unser Frau.) Just as any other hospice of the period, it was supposed to aid pilgrims and crusaders on their way across the alpine passes. Laurein and Proveis both appear to have been founded during the twelfth century, although Laurein is first mentioned in documents in 1233 and Proveis in 1274. Both were probably settled under sponsorship of the Counts of Eppan, also concerned with extending the settlement area of German speakers in the neighboring valley of Ulten. Proveis may well have been a hamlet of pastoral stations (*Schweigen*, see Altenstetter 1968: 50). It is notable in this connection that the first homesteads known from Unser Frau all paid tribute in cheese to the hospital; only Malgasott—located in a favorable, sunny spot—paid tribute in wheat (Gasser 1901: 84). By 1221, the hospital drew tribute from properties not only in Pizuvit and Malgasott, but also from Romance villages in the Anaunia, like Tret, Fondo, Malosco, Seio, Amblar, and Castelfondo, as well as from vineyards in Mais near Meran, Tisens, Nals, and Kaltern in the valley of the Etsch.

The Germanic incursion into the uplands of the Anaunia formed only part of the larger Germanic expansion into the area south of Bozen. This advance is marked by the construction of a chain of castles, first Hocheppan at the beginning of the twelfth century, extending southward to the confluence of Adige and Noce. There the castle of Kronmetz (Italian: Mezzocorona) guarded the entry to the Anaunia, while the castle of Visione on the Rocchetta surveyed the narrows of the Noce River and the passage across the bridge, which the Italians called Ponte Alpino and the Germans, Puntelpein. At St. Michael, now San Michele d'Adige, the lords of Eppan founded a German, Augustinian monastery as a religious stronghold alongside their military redoubts. After the decline of the Eppan, all these fortifications and religious foundations came into the possession of the Counts of Tyrol.

When the Germanic advance into the Upper Anaunia is plotted against the extension of the Germans in the valley, we see that two pincers crosscut important ecological relationships between the populations of the valley terraces and the highlands. The mountaineers used to send their livestock to winter in the meadows along the unregulated and swampy Adige River, whereas the valley settlers drove their cattle to the Anaunia, especially to Mt. Roén, in the summer. The founding of Unser Frau thus may be interpreted not merely as the creation of shelter and sustenance for weary pilgrims, but also as the establishment of a control point over mountain-to-valley pastoral trails. This interpretation is strengthened by the fact that the hospital at Senale soon drew tribute not only from its immediate vicinity, but from vineyards in the Adige valley, as well as from Romance villages in the Anaunia. We learn that the Counts of Tyrol, in the thirteenth century, settled German speakers in the village of Ruffré (formerly Fondoi), which guards the transit from the Anaunia over the Mendel down to Kaltern. This conjunction of historical events further explains why Unser Frau was frequently locked in bitter disputes with Fondo over rights to forest and pasture grounds located in the intervening zone, now occupied by St. Felix and Tret. The area around St. Felix, then called Caseid, is first mentioned in 1233 when one Vitus de Vasio is said to have possessed some "*case coloniche,*" huts of *coloni,* or dependent cultivators, "above Tret in the place called Caseid [Inama 1905: 48]." This Vitus of Vasio —the Vasio castle lies between Malosco and Brez—was probably a dependent of the Counts of Eppan, who erected the castle. Cultivation in the vicinity of Tret is first mentioned in 1294 (Inama 1931: 73).

From the end of the thirteenth century, Unser Frau, Fondo, and the intervening localities were continuously in disagreement over rights to woods and pasture grounds. In 1342, Unser Frau and Caseid disputed their rights over the mountain of Malgasott, that is, the zone of wood and pasture lying between Caseid and Unser Frau proper. It is significant that the settlement involved not only representatives from the neighborhoods of Untere and Obere (German: Bischofseyd), but also three witnesses from Fondo (German: Pfund), who would hardly have participated in the case if the interests of the Romance town had not been in question. The quarrel broke out anew in 1495 (fanned by the lord of Castelfondo), in 1580, and once more as late as 1860. Moreover, a lawsuit between Unser Frau and Fondo was fought over the area called Tullcamp— probably now the Felixer alm (high pasture)—and adjudicated in 1405. Fondo yielded the area to Unser Frau for wood and pasture, in return for a payment of empheteutic rent of sixteen Meran dinars annually, the contract to be renewed every twenty-nine years. This dispute flared again in 1556. Finally, Fondo and Unser Frau disputed access to Salomp, the mountainside lying between Caseid (St. Felix) and Tret; starting in the fourteenth century the arbitrations were repeated in 1405, 1444, and 1530 (Inama 1931: 28, 79, 89).

These repeated adjudications strengthen the probability that both the monastery near the Gampen Pass and the town located where the Novella breaks out of its transversal valley into the upland basin were vying over control of a frontier zone with scant prior settlement and generally undefined boundaries. At the same time, Fondo and other villages in the Upper Anaunia may well have made use of the Gampen Pass to drive their livestock into the valley of the Adige, rather than take them over the dangerous Mendel Pass. In that case, rights to pasture on the way would have been of double importance to the Romance settlers, and the closing of that route by the monastery at Unser Frau would have fanned political conflicts in and around the Anaunia.

In 1320 the lords of Arz, Malosco, Cavanno (Cavareno?), Raina, and Cloz formed a compact to oust the German overlords from Kaltern. The records say:

. . . they, the Nonsbergers, possess properties in Kaltern and bear German overlordship there unwillingly, for if the Romance ruled there, their condition would improve, therefore they wanted to kill Heinrich von Rottenburg [quoted in Stolz 1927, Vol. II: 72].

Otlin of Raina, one of the conspirators, confessed under torture that he had been hired by some people to burn Tret and Fondo and to assassinate several persons in Kaltern, "because they believe that they would be better off, if Italians ruled in Kaltern [Ausserer 1899; fn. 2, p. 101]." It also seems more than coincidental that, in 1321, immediately after the termination of the conspiracy, control of Unser Frau was transferred to the Augustinian monastery at Gries near Bozen, thus tying the mountain way station more closely to the growing Germanic domain in the valleys.

Tret is first mentioned by that name in the document of 1320. The name itself may either be of Romance derivation, from *tres*, a field of abundant growth near a mountain hut, or from the Germanic *trat*, or *tratten*, a field kept fallow for livestock grazing (Prati 1919: 264; Tarneller 1911: 463). Local folklore has it that the first settlers in Tret were horse tenders, whose occupation labeled the neighborhood of Cavallaia, still in existence. A house in Tret dates back to 1650. The little church of the community, dedicated to Santa Anna, was built in the mid-1700s and enlarged several times thereafter. From the beginning, Tret was politically, juridically, and ecclesiastically an integral part of Fondo: *inter montania de Tret, plebis Pfundi*, a document of 1394 calls it (Inama 1931: 120).

Various authors have fiercely debated, on nationalist grounds, whether the original population of Unser Frau and St. Felix were Romance or not. The argument for a strong Romance presence in this frontier zone is not particularly well founded; but conclusive evidence either for or against this position is scant. As we have seen, the first mention of Unser Frau, in 1184, is by its Romance name Senale, as *Eccl. s. Marie de Senali*. *S. Maria ex Silua*—the Latin counterpart to *Unser Frau im Wald*—appears in 1298, but in 1296 we also find *Ecclesia S. Maria de Senale de Valdo*, the church of Saint Mary in the Forest—hence the designation of *Waldner* for the inhabitants; and in 1325, we have *Unser Frawen in dem Wald*.

Senale was the original designation of the area near the church, the *gotzhaus gemain*, the commune of the house of God. Later the name applied to four neighborhoods, designated respectively as Obere (Upper), Malgasott, Untere (Lower), and Caseid (St. Felix). Obere contained, in all likelihood, the original nucleus of settlement.

Homestead Urban, one of the oldest homesteads in the area, may well have been the first one settled: it was once called Pizuvit (from *picetum*, pine forest), a name later Germanized to Bischofseyd. Still later, the name Bischofseyd was extended to the whole neighborhood of Obere. Table 2, which lists the dates of first documentary mention for the "traditional" homesteads of Senale and St. Felix, suggests that settlement had taken firm roots in Obere and Untere by the mid-fourteenth century, with only a few homesteads scattered along the periphery in Malgasott and Caseid. Malgasott and Caseid, in turn, seem to have received substantial settlement by the fifteenth century. The sixteenth century then witnessed intensification of settlement in all neighborhoods (see Table 2).

Malgasott is first so named in a document of 1318. Untere (also referred to as Masi Nigri) politically belonged with Unser Frau, but ecclesiastically to the parish of Castelfondo until 1773: it is first mentioned as *in Senale, ubi dicitur ad inferiores* in 1342; in 1352 *a li Niderari in Senale*. Caseid, first so named in 1342, initially belonged—like Untere—politically to Unser Frau, but ecclesiastically to Fondo. It became independent of the Fondo parish only in 1723, when it achieved the status of a separate curacy. Its name changed to St. Felix in 1742, although as late as 1772 we still read of the *regolani della communità Caseidra,* assemblymen of Caseid community. It became politically independent of Unser Frau/Senale in 1864. Juridically, all of Senale—Obere, Malgasott, Untere, and Caseid—belonged to Castelfondo, together with the Romance settlements Melango and Dovena on the right side of the Novella. Moreover, the different neighborhoods possessed different charters regulating their access to forest and pasture. Obere had its pastures on the Laugen (Monte Luc); Malgasott had use of Malgasott Mountain, although their neighbors in Caseid—who were assigned all land lying to the southeast of the Mühlbach, the old mill stream (*da mans zu Teitsch nenet Mihlbach* says the document of 1342)—continually disputed their precedence. Untere used the mountains above Castelfondo; and Caseid had access to Salòmp, over which it fought with Romance Tret. These neighborhoods appear to have had independent assemblies, though earliest documentary evidence is dated 1776, when a separate assembly in Obere enacted a charter for their pastures on the Laugen.

In 1759, we hear of *Rigler der loblichen Gemein in der teutschen Gegend bei St. Felix im Wald* (members of the assembly of the honorable community in the German region of St. Felix in the Forest); in 1772, as we mentioned above, we hear of *regolani della com-*

munità Caseidra (members of Caseid community). St. Felix, one of the four constituent neighborhoods of Unser Frau—each having its own council—became an independent commune in 1864. Together with Unser Frau, St. Felix was juridically subject to Castelfondo until 1827; thereafter to Fondo. Its ecclesiastic loyalties were long divided between Unser Frau and Fondo. The cemetery of St. Felix still lies within the jurisdiction of Unser Frau, at Sankt Christoph on the edge of Malgasott, but its priests were long drawn from Fondo. In 1723 St. Felix became a separate curacy, subject to the Fondo deanery: the first priests were Italians, but after 1800 the priests generally were German-speaking Tyrolese. Their position in the community is symbolized by their title, often mentioned in documents as *der Berger Herr*. The Felixers are often referred to as the *Berger,* men of the mountain, who live *im Berg,* as opposed to the *Walder,* the forest men of Unser Frau, who live *im Wald,* in the forest. *Der Berger Herr* is the lord of the mountain men. The church of St. Felix was consecrated, in 1742, to St. Felix, a saint martyred in Valle Gardumi, near Mori. Both Unser Frau and St. Felix formed part of the province of Trento until 1948, when they were transferred to the province of Bozen.

The conjunction of Germanic and Romance spheres of influence in the Upper Anaunia permits some measure of insight into the relations between overlords and peasantry. In its dealings with the Romance areas, Unser Frau followed the usual Romance custom—in turn codified in Longobard law—according to which the *colonus* was both personally free, and free to move if the property he worked was sold to another master. He was obligated, however, to furnish stipulated annual rentals, contracts renewable first every twenty-nine years, and later every nineteen years upon the additional payment of a pound of pepper. Contracts were confirmed by handshake, *per tactum manus,* the way contracts are sealed to this day. We have also seen that the inhabitants of communes owed services to overlords, as illustrated by the services granted by Fondo to the lords of Castelfondo. Similarly, the German settlers were obligated to render payments to the ecclesiastical establishment at Unser Frau, as overlord. In return the settler received hereditary rights to occupancy, against payment of rent; *secundum usum, jus et consuetudinem de Ultimis et ecclesie predicte sancte Marie de Senallo* (according to the usage, law and custom of Ulten and the above-mentioned church of St. Mary of Senale), it says in a document of 1396, in which the head of the monastic establishment at Gries grants to Nikolaus of Obkirch the homestead of Gasteig.

FIRST DOCUMENTED MENTION OF HOMESTEAD

Name of homestead	Century			
	14th	15th	16th	17th

Unser Frau

Obere

Name of homestead	14th	15th	16th	17th
Gori Weiss	1329			
Urban, formerly Pizuvit	1342			
Gasteig	1342			
In Traten	1342			
Marschalk	1342			
Pfarrhof	1396			
Obkirch			1524	
Jager			1524	
Hüttl			1576	
Langes			1576	
Rotnacker (¼ of Weiss)				1694

Untere

Ban Bruggen	1342			
In der Wies	1352			
Stricker (Orthof)	1342			
Büchler, Pichler			1542	

Malgasott

Klammer	1318			
Häusl	1342			
Bacher	1396			
Mair		1425		
Egger		1496		
Kindler			1519	
Messnergut			1524	
Ausserer			1524	
Wirt im Wald			1576	
Holzgut			1524	
Leitner			1524	

[a] From Gasser (1906) and Tarneller (1910–1911).

Name of homestead	Century			
	14th	15th	16th	17th

St. Felix

Name of homestead	14th	15th	16th	17th
Schwarzviertel				
Brugg	1342			
Lochmann		1427		
Larch			1524	
Bairisches Viertel				
Klamm		1476	1576	
Pfeifer		1476		
Grueb			1576	
Jordan			1598	
Lind			1599	
Tal			1599(?)	
Unterberg				
Kofler		1423		
Erspam			1524	
Greit			1539	
Rast		1427		
Nidrist			1579	
Nuiwirt			1546	
Blasing			1579	
Grill		1454		
Brunn			1516	
Waldner			1524	
Oberberg				
Rain		1496		
Sandegg, Tratner				
Rohregg	1364			
Balsern		1430		
S'Egg			1524	
Traten				1622

After 1495, such documents follow Tyrolese law, in which unrestricted use rights, including right of sale, are granted against stipulated services to the overlord. Certain homesteads are always described as independent, *luteigen*, from which it may be concluded that there existed independent peasant proprietors at least since the mid-1300s. This was also true in the Romance area, where a stratum of independent peasant proprietors soon came to enjoy tax exemptions, as did nobility, and began to identify with impoverished members of the noble class. Noble crests on old peasant houses are to this day a hallmark of Nónes villages.

PEASANT REVOLTS

Working in concert, the Bishops of Trent and Counts of Tyrol succeeded in putting down the unruly and rebellious lords of the mountain, in favor of more centralized management from Trento. Yet centralization was soon to produce its own problems—the year 1390 witnessed the accession of a bishop whose efforts at greater centralization led him to abrogate numerous local privileges, and whose option for greater independence from the heavy-handed protection offered him by his legate, the Count of Tyrol, exacerbated the opposition of communes, nobles, and Count. In 1405, the southern Trentino, led by the lords of Caldonazzo and Castelbarco, transferred allegiance to Venice, and in 1407, the peasants of the Anaunia rose against the Bishop and his supporters in the valley. Led by the Cazuffi of Tuenno and having the tacit support of the Count, the peasants proceeded to burn the castles of Tuenno, St. Ipolito, and Altaguardia in the Val Bresimo. The Bishop was forced to remove hated officials, to confirm and extend local privileges, and to agree to raze the three castles.

In 1477, again, the peasants rose against the episcopal administration and burned down Castle Coredo. Both maladministration by self-willed lords in the employ of the Bishop and heavy fiscal exactions appear to be the immediate causes of the rebellions. In long-term perspective, however, the repeated rebellions were clearly reactions to episcopal efforts to draw increased dues from the ever-growing population, often in violation of locally established traditions. The same factors were at work again in 1525, when the Anaunia joined in the pan-Tyrolean peasant revolt. The revolt began when German peasants from Lana, Tisens, and Nals attacked the pilgrimage center at Unser Frau and laid siege to Castelfondo. Once again, Nónes peasants all along the left bank of the Noce moved upon Trento, burning castles and attacking centers of the episcopal administration as they went. The main reason behind these uprisings was economic abuse; religious considerations were of little importance. Everywhere episcopal officials were deposed, and peasant officers elected in their stead. By the end of August 1525, however, the siege of Trento was lifted, and the peasant rebels were pursued by the Bishop's victorious troops. The rebellious communes of the Upper Anaunia had to surrender hostages and yield up the leaders of the rebellion at Revó; many rebels were executed. Forty-six peasants were punished with exile, among them Johann Grill of St. Felix and Anton Goasser (Geiser) of Unser Frau. Others were stripped of their civil rights to participate in the communal assembly (*Riegl*), such as Urban Kessler of the homestead Urban, N. Marschalk of homestead Marschalk, the brothers Kaspar and Heinrich Geiser, and Peter Schwaiger, all of Unser Frau (Ladurner 1867: 159).

THE IMPACT OF THE COUNTER-REFORMATION

With the peasant revolt quashed and the Counter-Reformation triumphant, Habsburg power extended its rule without resistance over the Anaunia, as over other parts of the Tyrol. Habsburg centralization put an end to the conflicts between the Counts of Tyrol and the Bishops of Trent, thus also ending the feuds of minor nobles in the valleys of Non and Sole. The power of the church was extended, under the imperial aegis by such personalities as Bernardo of Cles in the Anaunia, who became Bishop of Trento and whose extensive construction and reconstruction of local churches throughout the valleys furnished the outward expression of consolidation of the Trentine ecclesiastical establishment. The extended sessions of the Council of Trent (1545–47, 1551–52, 1562–63) not only initiated the ritual and intellectual activities of the Counter-Reformation, but also provided an important economic stimulus to the neighboring valleys on which the city depended for grain and meat. It is said that during the sessions, 3000–4000 head of cattle were marketed daily in the Trent markets (Zieger 1926: 116). It may have been at this time that the Anaunia came to be known as the "granary of Trent," *il granario di Trento,* as Giano Pirro Pincio expressed it in 1546 (Franch 1953: 33). Contemporary rhymes stress the prosperity of the Anaunia—the Tyrolese rhyme of 1558 states:

Auff Nonser Perg sind wol erpawen	On Non Mountain
Vier und Zwainzig Schlosser	One can see twenty-four well-built
z'schawen,	castles,
Vierthalbhundert Dörfer fürwar	Two hundred villages for sooth
Und vier und Zwainzig Pfarren gar,	And even twenty-four parishes,
Hat Visch und Wildprät, Wein und	Possess fish and game, wine and
Korn,	wheat,
G'schickt und g'lert Leut sind daruff.	Handy and learned people are there.

What we know of Fondo during this period supports a picture of economic improvement. In 1517 Fondo was granted the privilege of holding an annual fair (on July 25th) and a weekly Saturday market; its population grew as the result of immigration, mainly from the rural areas of the Anaunia (Dambel, Brez, Dermulo, Val di Sole), but also from as far away as the Engadin (Inama 1931: 36). The enhanced quality of life in the town is symbolized by the cheerful, now almost illegible, verses painted on the wall of the old inn (now Hotel Bertagnolli), which read:

Ich bin ein Succkopf wohl erkannt	A well-known cabbage head am I
Herein Gäste und Gesellen	Enter guests and companions
Die Hüner und Eyer haben wöllen,	Who want chickens and eggs,
Die ich mit meinem Hintern ausgebruet	Which I have hatched with my arse,
Die sind für die Pfundser guet.	These are good for the people of Fondo.
Viele Küsse mir winken	Many kisses await me,
Den Wein will ich wie Nectar trinken.	Wine I will drink like nectar.

In spite of this general economic impetus, rural life cannot have been completely free from turmoil. Economic difficulties again became apparent at the end of the sixteenth century, and toward 1600 the Anaunia was beset by the expanding witchcraft craze, which was affecting much of mountain Europe (Trevor-Roper 1967: 102). It is noteworthy that, although the craze began in the Pyrenees and in the old region of Catholic heresies in southern France as early as the fifteenth century, it did not take root in the Trentino until the sixteenth century. Precise data are lacking, but one is tempted to relate this spread to the general disruption of social ties in the area, a disruption occasioned first by commercial and mining capitalism, together with the peasant revolt, and later by the reorientation of life under the Counter-Reformation. The phenomenon reached Fiemme in 1501–05, Lüsen in 1548, Velturno in 1550, Anteselva in 1592, Sexten 1595, and the Anaunia in 1612–15 (Zieger 1926: 124).

For some time, renewed economic stagnation was masked by the increased investment in the bureaucratic apparatus of the Habsburg Empire; this proved to be of great importance to numerous noble families in the Anaunia, many of whose members filled important posts in both the secular and ecclesiastical structure. This was true not only of German families like the Thun, who had remained loyal to the emperor throughout the vicissitudes of the peasant wars, but also of many Romance families. The Counter-Reformation favored the entry of Italian Catholics into the bureaucratic ranks, frequently at the expense of the less reliable, German-speaking nobility. Thus, families like the Cazuffi of Tuenno, the Mollaro of Mollaro, the Barbacovi of Taio, the Concini of Malgolo, the Manincor of Casez, the Genetti of Castelfondo, and the Inama of Fondo entered the ranks of the imperial nobility, developing ties and influence that reached far beyond their narrow home base. Other nobles, however, fell into poverty, or moved into local professional positions, often combining their small incomes with some agriculture.

During the sixteenth and seventeenth centuries, many peasants were able to buy proprietary rights to their land and to convert their status from renter, whether temporary or hereditary, to outright owner. Thus, we know that most of the homesteads in Unser Frau and St. Felix became free homesteads, though their owners agreed to pay a tithe annually, at harvest time, to the church. One reason for the easy change from rental to ownership was the continuing decline of monetary values, coupled with the maintenance of fixed rentals, which undoubtedly made homesteads less attractive to their former secular or ecclesiastical proprietors (Trafojer 1947: 59). For the bulk of the rural population, however, emigration proved an effective alternative to remaining on the increasingly overpopulated land.

As early as 1546, Pincio commented on the commercial orientation of the Nónes—*hanno del mercuriale assai* (they are sharp businessmen). Many became peddlers. Like the inhabitants of other Trentine valleys, others turned to crafts or commercial specializations. The Val di Sole was known for its coppersmiths and tinkers, the Anaunia proper for its chimney sweeps. We know that the people of Tret became stonemasons and hired out as construction workers in other parts of the German-speaking Tyrol. The commercial proclivities of the Nónes became proverbial in the Trentino: *furbo come un Nónes*, "sharp as a Nónes," was at once snide and complimentary. Nónes and Solandri from the Val di Sole must have managed many a successful deal to inspire the ancient prayer: "Lord, liberate us from Nónes and Solandri—*A Nonensibus et Solandris libera nos Domins* (Schneller 1899, Vol. I: 90).

Emigration was to become ever more serious throughout the centuries, especially after the introduction of the potato into the valley in 1799 (Perini 1856: 198) caused a sharp rise in population growth in the second half of the nineteenth century. Whereas population in the area may have been as low as 8000 at the end of the fourteenth century—a document of the year 1387 lists 1200 hearths for the Anaunia and the Val di Sole together (Ladurner 1868: 346)—by the end of the nineteenth, we have estimates of 50,000 for the Anaunia and 19,700 for the Val di Sole; up to a fifth of the Solandri are said to have been working outside the valley (Schneller 1899, Vol. I: 63, 88). Such population pressures were to produce both a wholesale and permanent migration to the New World, especially to the United States, at the end of the nineteenth century, as well as to intensify seasonal migrations within Europe. Construction work on the expanding railroad network of central Europe became a special source of employment for the Nónes; many worked away from home, for a longer or shorter period of time, as *eisempòneri*, railroad workers (from the German *Eisenbahner*).

THE FRENCH INVASION

The revolutionary events of the late eighteenth century and the nineteenth century affected the Anaunia, though only tangentially. The French invasion of the Tyrol in 1796 and 1797 led to battles along the Adige and Eisack rivers, but it did little to change life in the Anaunian upland. As Christian Schneller stated:

> The Nonsbergers saw modern Franks in the guise of French troops then only when the French general Chevalier reached Cles on May 28th, 1797 with 108 footsoldiers, two dragoons and a small train, there proclaimed French liberty and equality to the Nonsbergers to the accompaniment of beating drums, and returned to Mezzolombardo on the following day, after consuming a plentiful lunch at Castle Brughiero. It is not recorded that the Nónes erected freedom trees with red caps and danced happily around them, as was the case in other places in Italy [Schneller 1899, Vol. I: 43–44].

Nónes and Solandri, did, however, take part in the rebellion of Andreas Hofer against the French invaders, despite the pessimistic prediction made by the Count of Arco when asked to head the rebellion in the Trentino:

> My countrymen are not as fit [for war] as the German peasantry and almost impossible to organize. They are generally patriotic, but at the

same time suspicious and argumentative. They have few rifles, even less ammunition and no money [Hirn 1909, Vol. I: 363].

Nevertheless, they responded to Hofer's call and participated in the holding action against the French division that advanced on Ala (May 2, 1809), rallied to the Trentine leader Dalponte, and took part in the siege of Trento against Peyri (October 5–10, 1809). When the siege proved abortive, many were made prisoners; others were shot when found with weapons in their hands. A few Trentini escaped to Spain, where they joined the Spanish guerillas in their fight against the Napoleonic invasion (Zieger 1926: 158). For a brief period (1810–15), under French hegemony, the five districts of Trento, Rovereto, Riva, Cles, and Bozen were brought together as a Department of Alto Adige, while the Vintschgau and the valleys of Eisack and Puster were grouped with the Kingdom of Bavaria. For a period of five years, the boundary line between the new Italian state created under Napoleonic aegis and the Bavarian kingdom passed along the Gampen to the north of Unser Frau. The defeat of the French armies once again moved the boundary line to the south.

POPULATION: ST. FELIX AND TRET

We have no population figures for St. Felix before the second half of the nineteenth century, but there are data, which make possible some educated guessing. The number of homesteads located in St. Felix, but paying dues to the church in Unser Frau, in 1524 was 12; in 1697 it was 29. By 1766, when the church in St. Felix had become independent from Unser Frau, the number of such homesteads had climbed to 31. Unfortunately, we do not know the number of homesteads that either were not required to pay any dues (the Pruggerhof seems to have been one of them), or that paid dues to the parish in Fondo; they seem to have been few, however. We have calculated the average size of family for the eighteenth century—the period before the introduction of the potato—at 9 if children not surviving past age 15 are included, and at 5 if they are not. Using this figure as a reasonable basis for calculation, we obtain the minimal values for St. Felix given in Table 3.

These figures allow us to estimate the total population of St. Felix for the sixteenth century, at 100–125 persons; in the seventeenth century, at 250–275; and by the end of the eighteenth century, at

Table 3
POPULATION ESTIMATES FOR ST. FELIX

| Date A.D. | Number of households | Minimum total population | |
		Including survivors to 15	Excluding survivors to 15
1524	12	108	60
1697	29	261	145
1766	31	279	155

275–300. The introduction of the potato in the early years of the nineteenth century produced a sharp upswing in population growth. Average family size, including survivors to age 15, rose from 8.45 in the eighteenth century to 8.68 in the nineteenth; if only survivors past age 15 are included, we get an increase from 5.03 in the eighteenth century to 6.66 in the nineteenth, a 13.24% increase in survivors. On the basis of these figures, we may estimate the natural increase in population at around 40 persons, from 300 in 1799 to 340 at the end of the nineteenth century, a figure that tallies nicely with the population figure for St. Felix cited by Staffler—348 for the year 1869.

Whereas it may be possible, as some Italian authors have alleged (for example, Donà, in Inama 1931: 144), that some of the original settlers in St. Felix were Romance speaking, this does not seem very probable. The register of tributaries from St. Felix to the church of Unser Frau of 1524 does not show a single Romance name. The most common surnames in St. Felix have always been Geiser (from German: Geis; Tyrolese goas, goat) and Kofler (from Tyrolese kofl, kefl, rock or mountain). "There is nothing here but goas und kefl," goats and rocks, say the local wits. By the end of the eighteenth century, however, Romance names begin to appear in local documents. A family named Cologna, from Castelfondo, buys land in St. Felix; Giambatta Cologna is married there in 1781. A Bertagnolli Treti, described as colonus in Gerait (laborer on Greiter homestead), marries a girl from Tisens in 1777. Early in the nineteenth century, two more families Bertagnolli, one originally from Cavareno, the other from Tret, establish themselves in St. Felix. By the time of our field work there were seven households headed by men named Bertagnolli or Cologna in the community. While their presence is generally accepted and they have intermarried freely with other villagers, their Italian origin is remembered.

The population figures for Tret are much less complete. We do have a figure for 1633, which indicates that there were eleven taxable households in the village at that time. Using a figure of eight or nine siblings per sibling set surviving past age 15 as a multiplier (see Table 16, Chapter X), we can estimate the population in the early seventeenth century, at 110–120. We estimate about 160 people for 1800, and roughly 200–250 for 1900. By 1960 there were 238 people. The most populous name groups in the community have always been Bertagnolli and Donà, and there is no reason to seek a Germanic origin for these. The Profaizers, however, attested in Tret at least since 1633, may well have had their origin in German-speaking Proveis, one of the four communities of the Anaunian *Deutschgegend*. Immigration by German speakers into Tret in any large number is, as we shall see, a phenomenon of the twentieth century.

THE IMPACT OF NATIONALISM

Although there was little overt political conflict between Nónes and Germans in the nineteenth century, ethnic relations between the two groups became increasingly antagonistic. The efforts of German nationalist educational associations in the Trentino proceeded apace, as similar Italian aspirations were realized in organizations of their own, especially after the establishment of the Kingdom of Italy.

After the French and Bavarian defeat and throughout the nineteenth century, the Anaunia followed the rhythm of development of the Trentino as a whole. The rural population in the main remained loyal to the Habsburgs, though the minor Nónes nobility and the professionals of the towns showed increasingly pro-Italian sympathies. Thus, the incursion of the Garibaldini into the Trentino in 1848 met with a sympathetic response in the Anaunia, and some volunteers joined the Italian cause. Rumors of an impending revolt in Fondo, in turn, mobilized the Germans in Unser Frau and St. Felix. They called upon a company of sharpshooters stationed at the Gampen Pass and, together with men from Kaltern, occupied Fondo in order to "nip in the bud" the revolutionary desires of the Romance speakers (Menghin 1897: 272; see also Zingerle 1877: 144).

From 1850 on, however, German-speaking communities in the Upper Anaunia came under increasing political pressure from the ever more nationalist Italian-speaking authorities in Fondo and Cles. This was especially evident in increased efforts to extend the use of Italian in offices and schools. To counter these pressures, the local priests of Proveis and Unser Frau sought outside financial support for schooling in German and for the development of local crafts that would render the German communities less dependent economically upon their Italian neighbors. Franz X. Mitterer, priest in Proveis from 1824 to 1899 and P. Ambros Steinegger from Unser Frau (1862–72) obtained funds first from the *Deutsche Schulverein,* and later from both the German-Austrian Alpine Association and the *Allgemeine Schulverein.* They also attempted to remove the four German-speaking communities from Trentine jurisdiction, but were unable to do so. Unser Frau and St. Felix did not become part of province Bozen until after World War II, in 1948. Nevertheless, they clung to Tyrolese Catholicism in order to maintain their ethnic identity against Italian encroachment. It is noteworthy that Michael Gamper, who was to organize the South Tyrolese catacomb schools during the Fascist regime, was a nephew of Mitterer and son of a man from Proveis, and was fully acquainted both with the way in which the priest from Proveis had organized outside help on behalf of his educational effort and with the pattern of education furnished by these schools (see Gamper, in Klebelsberg 1959: 170–172). Indeed, the practice of soliciting outside aid continues to this day. In 1966, an organization known as Silent Aid for South Tyrol (*Stille Hilfe Südtirol*) unveiled a kindergarten in St. Felix, which had been erected with $50,000 from outside contributions.

While the *Deutsche Schulverein* granted funds to underwrite a continued German-speaking presence in the *Deutschgegend,* other organizations experimented with the establishment of schools to further Germanification. Thus, in 1907, the *Tiroler Volksbund* established a German orphanage in Tret (the orphanage finally closed its doors in 1921 for lack of funds). Not to be outdone, the *Lega Nazionale* opened an Italian orphanage for Tret, which was put under the charge of the curate, Don Mattia Springhetti. This orphanage was closed in 1915, and the curate was arrested by the Austrian police, who charged him with irredentist activity. World War I also affected the communities in other ways. The Felixers gave their lives in battle, but the Trettners had to face divisions in their loyalty toward the Habsburg monarchy. Some men from Tret, taken as prisoners of war in Russia after Italy had declared war on Austria-Hungary, joined the Italian army; a few Trentini to this day consider it a point of pride to have served in the Imperial Hunters of the Austrian Army and continue to think of themselves as Tyrolese of Italian speech.

With the destruction of the Austro-Hungarian Empire, Unser Frau and St. Felix both came under intensified pressure from Italy.

The measures taken against German speakers elsewhere were applied to them and the reaction to these pressures were the same here as elsewhere. Both villages split into factions: there were the more Catholic oriented, who favored passive resistance to the Italian state; and those who longed for liberation by a strong National Socialist Germany. During the plebiscite, both communities opted for Germany, Unser Frau in larger percentages and with greater fanaticism, St. Felix in smaller percentages and with greater ambivalence. Many young men in both communities joined the German army and fought in Russia or North Africa; a few joined the S.S. When the Germans overran northern Italy after the collapse of the Kingdom of Italy, others joined the *Südtiroler Ordnungsdienst*, or paramilitary police, and made themselves useful to the Germans in the capture of downed Allied airmen, or in the impression into forced labor of some of their Italian neighbors in Tret. The activities of these eager policemen, in their green-brown shirts with the SOD armband, are not remembered very kindly in Tret. An entire family in St. Felix, with ties to the South Tyrolese anti-National Socialist movement, was arrested by the Gestapo and deported. In Tret, in turn, many sought to evade forced labor in Germany. Others, some seventeen men, formed a partisan band, connected in turn with a stronger group of partisans in Dovena, and ultimately joined with the larger partisan organization in the Anaunia, called Monteforte after the pseudonym of the Italian anti-Fascist Manlio Silvestre, shot by the Germans in 1944. The Tret partisans attempted one unsuccessful assault on the German army stronghold at Unser Frau.

Today no one in the two communities is eager to revive the memory of these incidents, perhaps for fear of kindling once again the internecine bitterness and hostility that surfaced so strongly during the war years. The years after World War II have, moreover, witnessed a rather extensive movement of German speakers into Tret. The first such family, that of the smith, had moved before World War II from St. Felix to Tret. Three households, displaced from Graun in the Vintschgau due to the construction of a dam and the development of a reservoir that flooded their holdings, arrived in Tret in 1949. A fifth household, coming from Platzers, bought land in Tret in 1956. A sixth family moved in from the Valley of the Puster to open a restaurant in 1958. Two landless households, one from Unser Frau and the other from St. Martin in Passeier, took up residence in 1957 and in the mid-1960s, respectively. More than 10% of all households in Tret thus are held by first-generation German speakers, offset only weakly by two Romance-speaking immigrant families from Revó and Mezzocorona in the Anaunia, who bought homesteads in Tret, one in the 1930s, the other in the 1960s. The first generation of German-speaking immigrants still cleaves strongly to German identity, but they have also adopted the patterns of social reciprocity and inheritance of their neighbors. The second generation has begun to intermarry with the Romance speakers of Tret, and is well on its way to acculturation.

Despite the immigration of Romance speakers into St. Felix 150 years ago, and that of German-speaking Tyrolese into Tret, the two communities have retained their respective identities to a surprising degree. In 1805 Jacopo Antonio Maffei wrote that in Unser Frau "there begins the use of the German language and mode of life, and the village is dispersed, and composed of homesteads." In St. Felix, "the German mode prevails [Maffei 1805: 94]." In 1876, the Nónes poet Bepo Sicher compared St. Felix and Tret: *"par todesč i parla sora, sot, envèze, l'Nónes sclet"* (German speech above, and Nónes, instead, below). The same holds today.

CHAPTER **VI**

Mountain
Husbandry

In tracing the flow of historical events that have affected the peoples of the Anaunia, we have emphasized the ebb and flow of local power relations, as these were influenced by competition between political rivals of wider political domains—first between the Counts of Tyrol and the Bishops of Trento, later between the protagonists of *Reich* and *Regno*. These events also have an ecological aspect: they succeeded each other within the same environment. In other words, they represent cultural facts that responded to the conditions of the environment and, in turn, determined the characteristics of that environment.

The terms of the dialogue between cultural heritage and local environment are complex. They are set by the interplay of local topography, flora and fauna, and climate with the cultural repertoire drawn on by the inhabitants of a particular area—the patterns of technology, organization, and ideology introduced by them. This cultural inventory is, in turn, the product of past ecological and cultural processes; the particular use to which the new environment will be put is in large part determined by a group's experiences in the past. The same environment could, for instance, present one set of opportunities and limitations to a group of pastoralists, and another to horticulturists. The modes of adaptation worked out by a society are as much the outcome of its past as of its present circumstances. The cultural patterns of Nónes and Tyrolese who came to inhabit the Upper Anaunia were distinct. While they have converged in response to a common environment, they have not come to be identical. Although the Nónes of Tret and the Tyrolese of St. Felix face common ecological problems, the two villages have maintained their cultural integrity.

At the same time, an appeal to past history is not sufficient explanation for the failure of these two villages to become culturally identical, to blend their distinctive traits into a common cultural tradition. This has indeed happened elsewhere—for instance— in Ruffré on the eastern periphery of the Upper Anaunia, which was established by German speakers, but is today wholly Romance. The decisive forces making for blending or continued ethnic separateness depend less upon different cultural origins, than upon the nature of the ongoing integration with the larger world.

Cultural processes in the villages, we have said, represent outcomes in the interplay of local ecological forces with forces emanating from outside the valley. We also know that, in this interplay, the relative strength and intensity of each set of forces does not remain the same. Over time, there has been a change in the nature and intensity of the forces originating outside the valley. Under the feudal system, the obligations imposed upon the population of the Anaunia from outside affected, in the main, the processes of distribution of peasant produce. With the growth of the capitalist market, outside demands began to involve peasant production as well as distribution. The spread of these demands on the Anaunia was slow; for a long time the area remained marginal to the economic processes of the lowlands, and production on peasant estates was carried on by marshaling local resources to ensure subsistence. Only recently have new alternatives arisen, allowing individuals to take up new options outside the villages. Our discussion reflects this changing order of significance. In this chapter, we are concerned primarily with the ways in which local resources are used to answer subsistence needs, to sustain domestic groups on estates carved from the sides of the mountains.

THE ECOLOGY OF MOUNTAIN AGRICULTURE

The Upper Anaunia, like other mountain areas, shows radical variations in both altitude and aspect of the land. The sudden alternation of valley floor and mountain wall has long appealed to mountain climbers, skiers, poets, and convalescents. The peasant who must wrest a living from the mountainside also has an appreciation of the land, though from a different perspective. He evaluates variations in topography and climate in terms of their ecological significance; and his appreciation of the land arises from his knowledge of how this variation can be used to provide the necessities of life.

Although there are plant species especially suited to high altitudes, as a general rule increasing altitude means the deterioration of climatic conditions suitable for cultigens. This is mainly due to variations in temperature: the higher the altitude, the colder the temperature at all times of the year. For most cultigens, cold springs and summers mean a retarded growth rate, so that the higher up crops are grown, the more time they require to mature. Increasing altitude also means a shorter growing season. Thus, as the amount of time needed for maturation is increased, the amount of time available for growing is decreased. Furthermore, altitude-induced climatic variation not only means that crops mature more slowly, but also that their growth is less satisfactory; hence, productivity per unit of land is low. The mountain cultivator must therefore anticipate an ever decreasing yield from his land as he moves uphill, until finally a limit is reached beyond which his crops will not grow at all.

Climatic conditions associated with high altitudes also affect the care and feeding of livestock. Foremost among the peasant's concerns must be the seasonal limitation on grazing. Whereas his animals may be turned out to feed in the summer, they must be maintained and fed in the barn during the long alpine winter. Therefore, a substantial portion of the wild grasses on which livestock feed must be mowed by the peasant and stored away for winter use. This requires a careful evaluation of available grasslands. Again, the peasant is faced with the fact that productivity per unit of land decreases with altitude. Low meadows will produce the most abundant harvests, a consideration that affects his choice of which land to harvest and which to use as pasture.

Within the limits of agricultural potential set by variations in

*"A substantial portion of the wild grasses on which livestock feed must
be mowed by the peasant and stored away for winter use." Raking hay, St. Felix.
The work group includes the owner of a homestead, his wife, his heir apparent,
and two daughters who have returned to the village to help in the harvest.*

altitude, other aspects of the terrain may restrict or prevent use
of the land. The character of the land's slope, exposure, soil, and
moisture must all be considered. Truly level land is at a decided
premium in alpine regions, and for the most part, the peasant must
be content with fields that exhibit some degree of slope. Increasing
slope makes the land more difficult to work and also increases the
danger of erosion. If steep enough, landslides are always a threat,
regardless of the use to which the land is put. With even moderate
slope, however, the natural tendency of soil to slide downhill is
aggravated in plowed areas. Grass, trees—any sort of ground cover
—reduce the likelihood of such erosion, but clearing the land for
the plow is, in any case, an invitation to gradual erosion or a sudden
landslide.

Even when altitude and slope are identical, fields may differ
markedly in their value to the peasant. In the Alps, considerable variation in the ability of the land to support cultigens results from variations in daily exposure to sunlight. Fields basking in the sun for most
of the day are far more productive than those that rest in the shadow
of a towering peak. Throughout the Alps this contrast is recognized
in high valleys, where village, plowland, and meadow are found on

the sunny slopes, while the shaded side of the valley remains forested, with only an occasional clearing for a meadow or pasture.

Still further variation in utility of the land depends on available moisture and the quality of the soil. Different parts of the Alps receive significantly different amounts of rainfall, depending on whether they are exposed to rain-bearing winds or lie in the lee of ranges, which relieve the air of its moisture. Even within major alpine regions, striking variations in rainfall are found, determined entirely by local conditions such as difference in the orientation of valleys and ridges and extremes in altitude variation. There are also variations in the ability of soils to hold water at high altitude. Rock covered by a thin layer of soil will not hold moisture, especially if the land has some degree of slope, while heavier soils in areas of abundant rainfall may retain excess water. In areas of little rainfall, mountain streams, which can be tapped with simple irrigation systems, may or may not be present; elsewhere, there will be parcels of land that lie directly downhill from a spring or lateral stretch of brook, where constant flooding may produce a worthless bog. Whether well watered or not, land parcels may contain soils of high or low quality, and they may be totally covered with soil, or fragmented by frequent rock outcroppings.

All variation in environmental characteristics of the land must be evaluated by the peasant in terms of the technology at his disposal. He must consider the specific requirements of the plants and animals he hopes to raise and also of the techniques and equipment available to him. The assignment of land parcels to their various uses is not, however, a straightforward decision arrived at simply by matching the growth requirements of each plant with the characteristics of each plot of ground. The range of fields set aside for crops will not be determined by the tolerance to altitude and aspect exhibited by each crop separately considered, but by their requirements as a group. Crop rotation, required in order to keep the land fertile, dictates that every plowed field must, in its turn, support each of the peasant's cultigens. Any parcel will therefore be eliminated as potential plowland if it fails to meet the requirements of any of the peasant's crops. This is especially notable with regard to altitude: how high up the mountainside plowed fields are to be found is a function of the least hardy crop the peasant plans to grow.

Parcels that make good plowland also make the best meadows, so that any land put under plow reduces available grassland. The peasant must therefore balance his requirements for crops against those for hay and divide his best lands accordingly. These fields will be at the lower end of the total range of lands available to the villagers and will also include the village itself. Collectively known as "home fields" (German: *Heimgrund*), these fields—and these fields only—are subjected to careful manuring. Human intervention here produces a major ecological difference between these manured fields and other land located on the mountain slope. Environmental variation is merely quantitative, growth conditions gradually worsening with altitude. Manuring of the home fields, however, substantially alters their productive potential, and introduces a qualitative distinction in the quantitative continuum.

Within the home-field area, however, the assignment of parcels to plowland or meadow is not an entirely arbitrary matter. While both require favorable slope, the degree of slope is less critical for meadow than for plowland. Meadow is continuously covered with vegetation, which retards erosion, whereas plowland is not only denuded much of the time, but because of plowing tends to be drier and to crumble more easily, and thus, is more likely to slide downhill. Recognizing this, each peasant is inclined to assign his more level fields to plowland and to leave those with greater slope as meadow. Furthermore, the subsistence cultivator must raise the full range of crops that he and his family will require throughout the year. He therefore must set aside a sufficient percentage of his home fields to raise all of the crops he will need. Once this need has been met, the remainder of his village lands can be used as grassland.

It is unlikely that the home-field meadows alone will provide enough fodder to keep the peasant's livestock throughout the entire year, so he must look to land above the village for additional grassland. He must consider how he can graze his animals during the warmer months and how much hay they will require during the winter. Some of the grassland, therefore, must be used as pasture and the remainder left as meadow to be harvested for winter feed. How grassland is apportioned between these two uses is by no means random. The number of swings of a scythe required to mow a field covered with dense, tall grass is not much greater than that required by the same size field thinly covered with short grass. Raking and transport time may be somewhat greater on the richer field, but the total effort expended per total return is, nevertheless, not appreciably greater. Peasants appreciate this fact and take it into consideration in dividing up their grassland: those lands that provide higher yields for less effort are used as meadow; those with the lower yields are set aside as pasture.

The highest yields of grass per unit of land derive, of course, from the home-field grassland; consequently, this land is used as meadow. Not only is its yield greater than other grasslands, but it lies close to the peasant's barn, keeping problems of transportation to a minimum. The hay provided by the home-field meadows is supplemented by the harvesting of high meadows (German: *Berg-wiesen*, Italian: *alpe*). Although there is considerable variation in performance, these fields provide considerably less bulk per land unit than the home-field grassland. Not only does the yield decrease the higher up the field, but the task of getting the harvest back to the barn becomes more time consuming and difficult. Finally, at some point on the mountainside, growth is so poor that cutting the grass is no longer deemed worth the effort.

In order to save grasses from village and high meadows for winter use, livestock cannot be permitted to graze there. Instead, a pattern has been evolved that permits the use of the most marginal growths of grass for summer grazing. As soon as snow begins to disappear in and around the villages in the spring and the land again begins to turn green, the animals are taken from their stalls and turned loose to graze during the day in the forests around the village. This is followed, as spring moves farther up the mountainside, by the division of the village herd into two parts, each with its own grazing area. The cattle and other animals required for day-to-day use, as dairy cows or draft animals are kept in the stall and continue to be grazed daily in forest and waste land in the immediate vicinity of the village. The remainder of the herd is taken to a high pasture (German: *Alm*, or *Alp*; Italian: *malga*) or range of pastures, where they remain for the entire summer, until snow threatens in the fall. The animals then return to the barn and once again are grazed in the surrounding forest and on village meadows (which by then have received a second cutting) until cold weather or snow forces the herds indoors. Then the long process of stall feeding begins, to continue until spring.

In addition to his strategies for getting the most from his fields and meadows, the peasant must consider himself in relation to the climatic conditions of his homeland. He must guard against the elements not only for his livestock, but for himself as well. A substantial dwelling is needed in which to live and to store all the things he will need to survive the winter. Once the snows begin to fall, there will be virtually no opportunity to augment his income; until spring he must depend on what he has set aside from the annual harvest.

The specific adaptation to alpine conditions of the European pattern of mixed agriculture—combining the harvesting of grains (and later, the potato) with animal husbandry—has served to maximize the population density of the high valleys in the Tyrol. Its typical features are a pattern of transhumance, which conserves the most productive terrain for plowland and meadow, and the intensive harvesting of grass from high meadows, which serves to maximize herd size. Without the hay harvested on the mountainside, meadows within the village would have to be expanded in order to maintain the herds; were the animals to be grazed there as well, a further increment of grassland would be required. Since villages have already pushed the available home-field land to the limits set by the ecology, this could be achieved only by a reduction in the amount of plowland. Yet, under conditions of subsistence agriculture, a reduction in plowland could not be achieved without a decrease in the population it supported.

Here, however, a cautionary note must be introduced: *ecological* limitations are being discussed, not absolute *environmental* limitations. Although there are absolute limits to which home fields can be pushed, the cultivated areas and tended meadows have not been pushed to these limits everywhere. It is the relative sequence of land categories that is repeated from community to community, not the absolute altitude range for any given category of land. The patterns of land use are imposed on the range of land available to the community. There has been a trend under conditions of subsistence-oriented production to extend cultivation ever farther up the mountain side, but this trend has sometimes been constrained by cultural and historic factors. The minimum harvest acceptable to peasant producers may be higher than the land can support; prior claims by lower-altitude communities who already exploit the land as pasture or forest may prevent its conversion to plowland and meadow; likewise political restrictions, which have nothing to do with the value of land to the cultivator, may be placed on expansion. Finally, given identical environmental conditions and demographic pressure, cultivation is more likely to be expanded into marginal areas where viable economic alternatives to farming are absent, than where such alternatives exist. Thus, the extent to which a given community approaches the environmental limits of cultivation depends not only on the man–land relationship, but also upon pressures emanating from the wider cultural and historical matrix in which each community is embedded.

Such variations are to be found not only between major alpine areas, but within quite constricted geographical contexts. Even within the Upper Anaunia, considerable variation in the absolute altitude of various categories of land are to be found.

THE ANNUAL CYCLE

The ecology of mountain agriculture not only limits the productivity of the land; it also regulates the annual round of human activities. In the high Alps, basic among the facts of life is the need to use the short growing season to best advantage—each household must provide for its yearly consumption during a productive season only six months long. True, there are elements in the daily routine that change little with the seasons: livestock must be cared for continuously, and there is no end to household chores. But the crops and hay required to support both man and beast must be sown, cultivated, and harvested during the summer months. This fact sets the rhythm of daily routine in the different seasons: leisurely in the winter, speeding up in the spring, unremitting in summer, and slowing again in the fall.

Through the ages, the mountaineers have sought ways of leveling out this seasonal imbalance, but with little success. Craft specialists, such as carpenters and blacksmiths, and innkeepers can ply their trades regardless of season, but—innkeepers apart—their skills are needed more often in the summer than at other times. Other craftsmen, most notably the stonemason, find their work seasonal: the stonemason requires the same fine, clear weather as the cultivator, so that his trade increases, rather than lessens, the activeness of the summer months. A few specialties, such as making rakes or fruit boxes, or, in the past, hobnails, can be carried on in the winter. Although they provide a welcome increment to household income, these pursuits are clearly supplementary and do not reduce the main burden of working the land and caring for the family herd.

The annual round of agricultural activity begins in the spring, as snow disappears from field and meadow (see Figure 5). By March, villagers cart manure from barn to meadow and, once the fields have thawed, they work to replace the soil that has eroded downhill during the previous season. This is done by hauling soil from the bottom of the field to the top, a process required annually on steeper fields, less frequently on those of lesser slope. Manure then can be distributed on the fields, and plowing can begin. The pressure of required

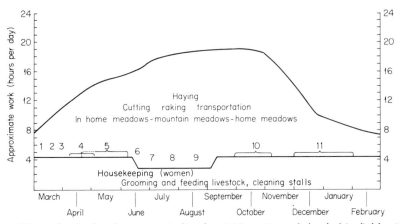

Figure 5. Cycle of major agricultural activities: (1) earth hauled in fields; (2) fields manured; (3) fields plowed; (4) spring planting; (5) home and mountain meadows cleared of debris; (6) and (7) potatoes hoed; (8) winter rye harvest; (9) winter wheat harvest; (10) potato harvest, plowing and sowing of winter grain; (11) miscellaneous tasks of maintenance and repair, logging.

tasks intensifies as the weather continues to grow warmer in April and May; spring planting and clearing meadows of accumulated debris take the most time. During June, with the onset of haying, the pitch of activity reaches its peak; it does not slacken again until October.

More than one villager has been heard to lament, "I live for my cattle," for the effort in cutting, turning, raking, transporting, and storing hay is greater than that devoted to all other agricultural activities combined. During the height of the season, everyone is expected to pitch in. The old people, who at all other times of the year are confined to their apartments, turn out to help with the raking, and even the youngest children are expected to do all the work of which they are physically capable. This season also puts to the test the filial responsibilities of those who have left the villages to take up work elsewhere. They are expected to temporarily set aside their own careers for the duration of haying, in order to assist their village relatives with rake or scythe.

The summer is truly a time of unrelenting toil. It is common for villagers to rise several hours before sunup, so that they can care for their animals, clean the stalls, and still be on their way to the fields by the first light of day. The women of the household stay behind until their daily chores are completed, but they hurry through them as fast as they can in order to join the men in the fields as

Manuring, Tret. In the background, the Laugenspitze/Monte Luc.

quickly as possible. Often one of the women remains behind to finish chores left undone by others and to cook the midday meal, which she carries to her family in the fields: no one can waste precious daylight hours traveling back and forth to eat. However, if the weather is fine and the villager feels that he is "on schedule" for the season, a short siesta is in order after lunch. Everyone simply stretches out on the freshly mowed fields under the warm, midday sun.

Soon the entire family is back at work. The men, armed with scythes, cut the hay. The women either turn hay with their pitchforks in fields cut the previous day, or rake dried hay into piles, preparatory to loading it on the wagon. Only at dusk does the family return home, often timing their arrival to coincide with the setting of the sun. The day's activities still entail unloading the hay wagon, caring for the animals, and completing essential, small tasks. It is well after dark before the family finally falls into bed; in five or six hours, all will be up again to begin another day.

Each household harvests its meadows in a well-defined order. Haying begins on meadows within the home-field area, because these lie at lower altitudes and come to maturity first. They also provide the highest yield for the least effort. These are mowed by mid-July, and the family then moves on to its mountain meadows, harvesting first those located at lower altitudes and following the ripening of

"Fields of potatoes require hoeing twice. The second hoeing requires a special horse-drawn plow—the plot is not only weeded, but plowed to leave a furrow between rows." View from Tret into the valley of Fondo; in the far distance to the right, the range of the Brenta.

the grass to ever higher altitudes. Many households have built sheds on the highest meadows, as temporary storage bins for their hay. The sheds are also sometimes used as sleeping quarters, for periods of up to a week, by men at work in the high meadows, as it may take a 4–5 hour walk from the village to reach them.

Mowing of the mountain meadows is completed some time in August. By this time, however, the grass on the home-field meadows has again ripened, and a second cutting is done. In villages farther down the mountainside, with a longer growing season, a third cutting is possible. But in Tret and St. Felix it will be October before all of the village meadows can even be cut a second time. Only in meadows recently converted from plowland, and in a few especially well-watered fields with better than average exposure, does the grass grow fast enough to permit a third cutting.

While haying is the main activity in the summer months, there are other tasks that must be worked into the schedule. Fields of potatoes, planted by mid-April, require hoeing twice, once near the middle of June and again around July 1. The first hoeing is done by hand; the second hoeing requires a special horse-drawn plow—

"Winter wheat is ready to harvest during August." Conjugal family harvesting with scythe and sickle, St. Felix.

the plot is not only weeded, but plowed to leave a furrow between rows, which drains off water in the event of heavy rains. Each of these tasks requires about a day of a family's time. Later in July the winter rye comes to maturity and requires a few days to harvest. The fields previously sown with rye are then plowed and usually either sown with oats or planted with cabbages and other garden crops. The oats are grown to be cut later in the fall, while still green, and fed green to the livestock, as a highly valued dietary supplement for dairy cattle. Winter wheat is ready to harvest during August, another task requiring several full days of effort. Since rain is likely at this time of year, and often interferes with getting in the wheat, the villagers harvest the grain as it matures, even if this means harvesting only a part of a given field at a time. Even so, the grain is often water soaked and useless for making bread; it then is used as feed for livestock. Soon after the wheat fields are harvested, they are plowed over to inhibit weed growth. In the spring they are planted with potatoes.

The harvesting of other crops, most especially potatoes, comes in October, after the haying is finished. When the potato crop is in, the potato fields are plowed and planted with winter rye and wheat. These two grains are rarely planted in the spring, and then only because the winter crop has failed or because some emergency prevented their planting in the fall. The growing season is too short to allow the cultivators in the Anaunia to count on a harvest of spring-planted grains.

Throughout the summer and fall, the grueling daily routine is broken only for the Sabbath, an occasional church holiday, or rain. Yet even on these days the animals require care, and the household routine must go on. Moreover, if any villager has fallen behind in his schedule, he will work in his fields on Sundays. This is especially likely if there has been a prolonged spell of bad weather.

The villagers live in constant fear that a long rainy spell will ruin a harvest of grain or hay, but they appreciate an occasional rainy day during the summer, not because it provides a day of rest, but because it offers a chance to tend to equipment and to put the barn in order. While the sun shines, a cultivator cannot relax his effort to harvest the maturing hay and crops. Always present is the fear that what is put off today may be impossible to do tomorrow because of bad weather. The goal during good weather is to get as much of the harvest as possible "under the roof." Because a man is tired when he returns home in the evening, and because he still has chores to do, unloading the day's harvest is done quickly, rather than with especial care. For the same reason, he is likely to continue using equipment in need of repair as long as it remains functional. We have even noted cases where a wagon or a plow that needed mending was simply set aside and neighbor's equipment borrowed, rather than waste any part of a sunny day in repair work. Thus, a rainy day finds the villager busy, stowing away crops in a proper fashion and in putting equipment back in order.

In the autumn myriad small tasks fall due. None is particularly difficult or time consuming by itself, but taken together they place yet another burden on the villager. Supplementary root crops must be harvested, and families owning fruit trees at lower altitudes must pick the fruit. One of the last crops to be harvested, some time after the first of November, is cabbage. While some cabbages are used fresh, most are shredded into a barrel and left to ripen into sauerkraut, one of winter's staples. Finally, before snow comes in the late fall, several trips are made into the forest to collect the winter supply of firewood and to gather fallen larch needles, which are used for litter in the stalls.

Threshing was once a task relegated to the winter months, but in the mid-1960s a lowlander with an ancient threshing machine inaugurated an annual comedy of frustration by arriving for a day or two in the middle of the fall harvest. The thresher—a marvelous contraption—not only threshes grain, but bales straw as well. Effort is saved by its use, but not time. Villagers arrive at appointed times, their wagons loaded with grain. Inevitably, however, there are delays—the machine is prone to breakdowns, and requires as much time to repair as to operate. Thus, villagers find themselves waiting in line long after their scheduled turns. Finally, the owner of the threshing machine moves on, without having served all his customers—he is scheduled to appear in a neighboring community and is already late for his appointments.

By November the pace begins to slacken. The cultivator now can arrange his schedule to his own convenience, rather than having it determined for him by the ripening grain and hay. There may still be some cabbages to pick or some root crops to dig up, but for the most part his work is now in barn or stall, storing away the harvest and caring for his animals. The sun now rises later and sets

Securing the winter supply of firewood, Tret.

earlier, and the villager sees little reason to do anything but follow its slower schedule. "The proper work for winter," one octogenarian informed us, "is to sit by the fire and see to it that it doesn't go out."

Yet even the winter has its special tasks. The rake-maker and the carpenter who builds fruit boxes now have time to devote to their specialties. Soon after the new year, each household butchers its single pig and processes the meat into *Bauernspeck* (smoked bacon) and sausages. Although most people today prefer store-bought clothing, older women still spend many winter hours sitting at their spinning wheels or with knitting needles in hand, turning out heavy, somber-colored woolen socks, trousers, and shirts. This is also the time when baskets and brooms are made from reeds gathered earlier in the year, when equipment is repaired, and when the interiors of buildings are refurbished and spruced up.

Winter is also a time for lumbering. Few villagers today find it worthwhile to cut from their private forests more firewood and lumber than the household needs, but there is lumbering work in other parts of the Tyrol and some villagers take advantage of these opportunities for employment. Some lumbermen even earn a bit of money in their home village, as each village council authorizes the cutting and sale of a measure of timber from the communal forests. The proceeds from the sale, after the woodcutters have been paid, goes toward the upkeep of the communal government; in St. Felix, indeed, the proceeds are expected to cover governmental costs completely.

Finally, some of the villagers try to get a head start on spring by hauling manure on sleds to the snow-covered meadows and fields so that it will be ready to spread when the spring thaw finally comes and the cycle begins anew.

Whereas there is a seasonal rise and fall in the requirements of the land, other activities demand the attention of the villagers throughout the year. For women, child care and domestic chores go on, regardless of season. Except in a few homes that have added automatic washing machines within the last few years, the family wash is done by hand, mostly outdoors at one of the communal fountains scattered throughout the village (Tret), or at the washing trough owned separately by each household (St. Felix). In summer, the drudgery of this task is somewhat relieved by fine weather, but in the winter the burden of outdoor washing is aggravated because women must work in the cold with their hands submerged in water only a few degrees above freezing. Indeed, we have seen women break a layer of ice to free the water in the washing trough. Village women also labor hard to maintain orderly and clean living quarters. The endless round of preparing meals, doing dishes, mending clothes, and cleaning house takes time and energy. Since not all households have running water, many women must carry all the water needed each day from fountain to kitchen. The floors in most homes are of unfinished wood and are kept gleaming white by the women of the house, who attack them at least once a week on their hands and knees, scrub brush in hand. Besides these chores, the women also tend the poultry and pigs, and help the men with the larger animals when necessary.

For the men, the care of livestock is the major year-round task. The jobs of cleaning the stalls, and grooming and feeding the animals require four or more hours each day, regardless of season. However, it is the one set of tasks somewhat lightened during the summer. Between the end of June and the middle of September animals not required daily by the household—including young cattle, sheep, and goats—are sent to the high pastures to graze under the eye of the village herdsmen. Even so, all the dairy cows and draft animals are kept in the barn. Since these make up the bulk of the village herd, their care and feeding continue to require the villagers' time and energy all summer long. This burden is felt most keenly in September and October when the hay harvest is in full swing, and the animals returning from the high pastures bring work in the barn to a maximum. In winter, when there is little else to do, however, the care of the household's animals is a welcome diversion.

Increasing emphasis on the production of milk for market has regularized monetary receipts, but has not relieved the pressure from uneven seasonal demands on household labor. The dairy needs a year-round supply of milk and pays suppliers by the month; thus, it provides each household with income throughout the year, rather than concentrated in the harvest season. The hay to feed the cattle, however, must still be harvested in the summer and autumn, and if anything, commercial milk production has increased the pressure of work during the summer months. The strategy of subsistence production dictated only that cattle survive the winter: stories are still often told of cattle that had to be literally carried from stall to pasture in the spring because of their weakened condition. Today, however, the villager must strive to keep his livestock healthy throughout the long winter, and he must, therefore, provide even more fodder for his animals than in the past. This has prompted the conversion of plow-

land to meadow and increased the competition for additional meadow; it has also increased the imbalance of the annual round of
activities.

THE ORGANIZATION OF VILLAGE RESOURCES

The communities found in the high valleys the length and
breadth of the Alps represent various cultural backgrounds. Furthermore, while the circumstances of their settlement and the details of
their subsequent interaction with the outside world have provided
each with a distinctive history, in every case they had had to evolve
social and economic institutions that could realize the ecological potential of the mountainous terrain. One finds effective use of an entire
alpine valley by a communal organization whose members hold all resources in common; by a division of a valley into two or more interdependent communities, each of which exploits some aspect of the
environment (for example, separate herding and agricultural communities); or by a series of independent households,each holding exclusive rights to a portion of each ecological zone. All these types are
known, either ethnographically or historically, in different alpine
locales.

In the Tyrol and the Trentino, high alpine valleys have been exploited by communities with access to the full range of the land's
potential: in its ecological relations one community is like the next,
so that mutual interdependence based on complementary economic
specialization is minimal. The population in smaller valleys is likely
to be organized politically as a single entity (German: *Gemeinde*,
Italian: *comune*), which will control all the land in the valley; larger
valleys are divided into several communities, each with a territory
consisting of a lateral section of the valley. Every slice of the valley
consists of land on each side of the stream that inevitably bisects it
longitudinally, up to the summit of the ridges that separate it from
neighboring valleys. Other mountain communities, however, are not
found in confined valleys, but are located rather high up on the
slopes of the major valleys.

The villages of the Upper Anaunia, lying along the northern
limits of the Val di Non, are of the latter type (see Figure 6). The high
land that forms this northern boundary is roughly in the shape of an
inverted "V," mirroring in its form the bend in the Noce River as it
flows into Lake Cles from the southwest and out again almost directly
south. At Lake Cles the altitude is only 525 meters—the banks of the

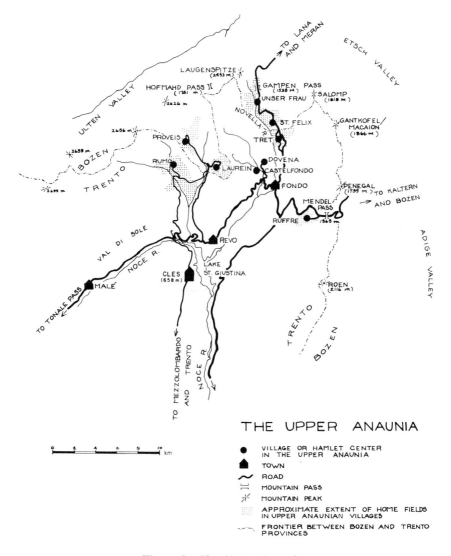

Figure 6. The Upper Anaunia.

lake are covered with grapevines and neighboring land supports many orchards. Yet the land climbs rapidly to the north, to heights of 1700–2700 meters, marking the crest of the valley lying within 10–17 kilometers of the lake. Fondo, which serves as a market and communications center for much of the Upper Anaunia, lies near the eastern arm of the "V," at 988 meters. The villages of Tret, St. Felix, and Unser Frau lie on an axis running north from Fondo to the

Gampen Pass, or Passo Palade, which provides a link with the Etsch Valley.

Towering over the Gampen Pass to the west is the major topographical feature of the northern end of the Anaunia, the Laugenspitze, or Monte Luco, at 2433 meters. Together, pass and mountain form the point of the "V." Unser Frau, lying immediately below, holds land on the mountainside as well as along the eastern arm of the "V." The lands exploited by St. Felix and Tret also lie on this arm, climbing to a thousand meters above the Etsch Valley. The Novella, a stream that originates high up on the Laugenspitze and flows generally southward, has cut a deep gorge along most of its length; this forms a convenient boundary between the three villages—Tret, St. Felix, and Unser Frau—and their neighbors to the west. The other villages of the Upper Anaunia are nestled into the western arm of the "V," each exploiting land above and around the village, up to the crests that separate the Val di Non from the Ulten Valley. An intricate network of transverse ridges and mountain streams, some flowing through veritable canyons, cut through the western arm of the "V" and provide the communities established there with convenient "natural" boundaries.

Each village controls lands distributed over altitudes extending from 800 to over 2000 meters (see Figure 7). Land use within this range of altitudes follows the general alpine pattern. The most favorably situated lands, relatively level and lying near the lower limits of the range, are in each case devoted to settlement, plowland, and meadow; fields at higher altitudes are devoted to forest and high meadows, and higher still stretch the summer pastures. While the altitudinal sequence of lands is the same from village to village, there are interesting variations in the absolute altitude of holdings. Unser Frau, at one extreme, has no land held by the community below about 1000 meters; in Castelfondo almost all of the plowland and meadow lies below that altitude. Land lying at 1000–1500 meters in Castelfondo is used as mountain meadow, or has been left forested, whereas the same altitude range supports plowland and meadow in Unser Frau. Similarly, the location of summer pastures varies from village to village. The Fondo pasture, also used by Tret, is situated at an altitude of only 1470 meters, whereas those of Unser Frau, Rumo, and St. Felix lie at 1700–2200 meters. Clearly we are dealing here with ecologically determined land categories: both environmental and cultural factors must be considered among their determinants.

The minimal resource unit in the villages of the Upper Anaunia, as in the Trentino and the Tyrol at large, is the *estate* (German: *Hof,*

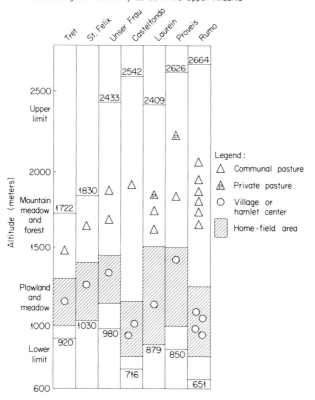

Figure 7. Altitude range of community lands in the Upper Anaunia.

Italian: *maso*). Each estate ideally provides a single household, or domestic unit, with access to the full range of ecological variation of the village through a combination of exclusive rights to land parcels held as an integral part of the estate, and rights to the use of lands held in common by the community.

The focal point of the estate is a house–barn complex, which furnishes each domestic unit with living quarters, storage facilities for its equipment and harvest, and shelter for its animals. Land in and around the village that has been cleared of forest is divided into parcels of plowland and meadow and is individually owned, as are the high meadows, which appear as clearings in the forest above the village. Ideally every estate should include land in each of these categories, and upon this land the domestic unit hopes to grow all the crops it will consume, as well as all of the fodder that will be required to winter its animals. Firewood for heating and cooking, and lumber for the construction and repair of buildings and equipment is obtained either from the communal forest, or from parcels of forest-land held as an integral part of the estate. Although private pasturage is common in other alpine areas, very little grazing is carried out privately in the Anaunia; instead each estate has the right to graze its animals on land held by the community as a whole. In the spring and fall, the entire village herd is turned out in the communal forest, and for the duration of the summer all but a few dairy cows are maintained by a herdsman on a high pasture owned by the village. Below we will examine the permutations of social and property relations within these units, but for the present our purposes will be served by considering some of the characteristics shared by the estates.

Each domestic unit may properly be regarded as an independent entity. Although the estate is constrained by social and economic ties to other households and by relations with village and regional or national institutions, decisions regarding the disposition of its resources are the exclusive prerogative of the domestic unit that holds it. How land and buildings are to be used, and the extent to which rights to the use of communal resources are to be exercised, are matters for the household to decide, as is the disposition of goods produced on the estate. In the absence of a market, opportunities to accumulate capital are few, and the estate manager rarely is able to expand the scale of his operation. Some rearrangement of land parcels may follow from the provisions of a dowry or will, and an occasional piece of land—or even an entire estate—may be sold, but the manager expects his land resources to remain constant throughout his life.

Moreover, the technology of mountain agriculture is simple: the peasant controls most of the skills needed to fashion his tools and materials for their manufacture are at hand. In the rare instances when he does require assistance, he can usually obtain it within the village through labor exchange or payment in kind. Except for the occasional purchase or repair of a metal tool, the village is self-sufficient in the manufacture of its technological inventory.

With estate size constant and with raw materials at hand, labor remains the strategic variable in the equation of household production. Through the application of labor, raw materials become the instruments of production, land yields the produce that is the wherewithal of the peasant's existence, and meadows grow the feed needed by the herd. Within the limits of each estate's ecological potential, production varies as labor invested varies. However, returns will begin to drop at some point—the peasant will experience diminishing returns per unit of labor expended. How far beyond this point the peasant is willing to go will depend on what is necessary for meeting the subsistence requirements of his household: as the size of the estate decreases, the number of labor units invested per unit of land tends to increase. The peasant will stop investing labor only when his requirements are being met. In the Alpine context this labor investment is perhaps most notable with regard to the maintenance of livestock: we already have seen how the mountain peasant makes up for a lack of home-field meadows by an intensive labor investment on low-yield mountain meadows.

While the Tyrolese *Hof* and the Romance *maso* are equivalent in that they provide access to parallel sets of resources, they are not identical in the details of these arrangements, nor in physical and political relations among estates. Tyrolese upland communities typically consist of dispersed households, or *Einzelhöfe*, as they are called in German. Living quarters, barn, and workshop are most frequently contained under a single roof, and each building is separated from the others by an expanse of field and meadow. Ideally, each building and the surrounding land belongs to a single household. Even today the impression gained by viewing, for example, St. Felix from the mountainside above is that the German villages of the Upper Anaunia approximate this ideal quite closely (see Figure 8). In St. Felix one sees five structures clustered around the church, and a half-dozen newer structures, some not yet completed, lining the new road through the village; but the remainder of the sixty-one inhabited buildings are scattered more or less uniformly throughout the home-field area of the village.

With closer inspection and inquiries about who lives in the various dwellings and owns which parcels of land, the picture changes somewhat. A careful observer on the mountainside might notice that what at first glance appeared to be a single building complex may be a cluster of two, three, or even four sets of structures. A visit to one of these complexes would confirm that there are indeed several distinct households living there. Further inquiries would reveal that certain structures house more than one family. A census taken in 1965 showed that the village's sixty-one inhabited buildings contained a total of seventy-four apartments.

Taken together, the plots of land owned by each of the households within a building complex comprise all, or nearly all, of the land lying immediately at hand, although within this area each house-

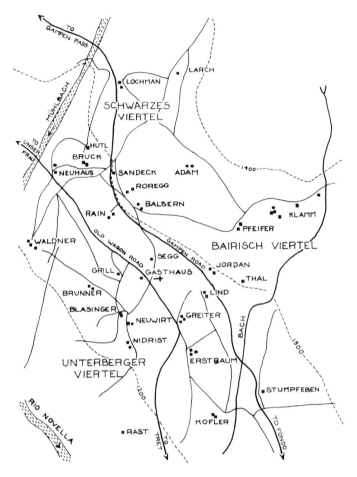

Figure 8.   Map of St. Felix.

hold's plots will be scattered and intermingled. It is quite evident
that this dispersal of landholdings is the result of divisions of property within a single line of descent. With few exceptions, the
genealogies of the present inhabitants of a building complex reveal
a common ancestor who held as a single estate land that several
descendants now operate individually. The few exceptions have resulted from the sale of an estate whose present owners or their
ancestors purchased the holding from a relative. It may also happen
that a parcel of land within the area belongs to a neighboring estate
that is not a part of the building complex, and if such is the case,
one can be sure that the details of its loss through sale or dowry are
well known to the householders.

The picture of farmsteads scattered over the landscape presented by German upland villages contrasts markedly with the Trentine pattern of nucleated settlement. Whereas nearly every Romance village includes a few dwellings set off by themselves, most apartments, barns, and workshops are clustered together into a dense mass of structures. Indeed, each structure typically houses a number of apartments whose inhabitants bear no necessary relationship to one another. Moreover, since the south wall of one structure may well form the north wall of the next, the buildings extend in one continuous mass, interrupted here and there by a passageway or kitchen garden, until the village abruptly ends and the fields begin. The plowland and meadow surrounding each village is divided into numerous plots, and each household has parcels of land scattered throughout the home-field area.

While this pattern is virtually unchanging in the lowlands of the major valleys, it varies somewhat in smaller valleys and in the uplands. Here, scattered homesteads are more common, and villages often encompass two, three, or more hamlets, rather than a single population cluster. Tret, and the other Romance villages of the Upper Anaunia show this modified pattern. The commune of Rumo consists of a series of four hamlets strung along a stream, the Torrente Lavazzé, for about 3 kilometers, with a number of single holdings interspersed between and above them. Castelfondo contains two neighborhoods, one lending its name to the *comune,* the other, Dovena, located a kilometer or so away at a higher altitude. Again, a number of isolated households are to be found scattered around the two population clusters.

Tret is perhaps the least nucleated of the Upper Anaunian villages (see Figure 9). Politically it is but a neighborhood, or ward (*frazione*), of the *comune* of Fondo, itself a highly nucleated town of over 1200 inhabitants. One large cluster, consisting of seven apartments and associated barns and workshops, as well as a store, fronts on the small village square. Other buildings are either located on the square as well, or are grouped nearby along the old wagon trail running through the village between Fondo and St. Felix. A smaller but reputedly older hamlet, Cavallaia, lies a few hundred meters uphill, and a number of other apartments, either singly or in small groupings, are scattered about in close proximity to these two clusters. There are also a few new constructions along the Gampen road. However, all buildings are within 300 meters of the church, and, whereas a few land parcels do lie between the outlying structures and the main hamlets, the bulk of the village home fields lies outside the

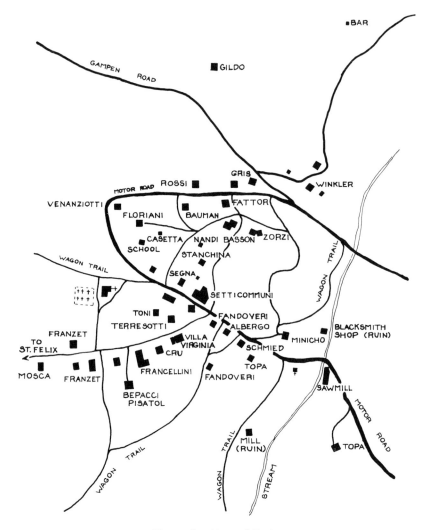

Figure 9. Map of Tret.

built-up area. A few domestic units hold reasonably concentrated units of land, but in most cases, the holdings are made up of numerous, scattered plots.

Separated by deep, stream-cut gorges, the home fields of each community are discrete: only rarely does a citizen of one village own parcels of meadow or plowland in a neighboring community. This holds equally true for adjacent villages of the same or different ethnic identity. However, higher up the mountainside, the gorges disappear, and the political boundaries between communities do not have the firm, "natural" basis found at lower altitudes. Villagers have appropriated land suitable for mountain meadows in a helter-skelter fashion that has juxtaposed parcels owned by one village with those of its neighbors. Thus, an estate may very well hold parcels of mountain meadow that lie within the confines of a neighboring commune.

Villagers tend to secure land—mainly forest and pasture land—for their own exclusive use above the village, although rights of access to such land vary among the villages, most notably between German and Romance communities. In the Romance communities virtually all forest land is held in common, but in the German villages substantial portions of the forests are privately owned. While the German estate expects to own sufficient forestland to supply all the wood it needs, in the Romance villages households depend on allotments from the communal forests. However, in the Upper Anaunia there are communal forests in all the German villages but Unser Frau, and villagers with small holdings depend upon wood from these forests. Moreover, lumbering carried out seasonally in the communal forests finances—at least in part—the operation of local government, and traditionally the communal forests have been used for seasonal grazing of livestock, a practice recently modified in accordance with state-sponsored reforestation policies. Pasture land, too, in the Upper Anaunia is held communally; animals from other communities may graze on these pastures, but such access is regulated by the commune owning the land.

Above, we looked closely at the relationship between man and nature in the Upper Anaunia under conditions of subsistence agriculture and we briefly discussed how resources are distributed among estates and communes. In subsequent chapters we will return to the question of how individual members of the community gain access to village resources. Before doing so, however, we wish to examine ecological relations in the Upper Anaunia that have unfolded as part of the cultural process of the Tyrol as a whole.

THE LIMITS OF EXPANSION

In tracing the history of the Anaunia we have seen how a Romance-speaking population pushed into the valley in early times, utilizing the Upper Anaunia proper for pastureland until about the eleventh century. Similarly, the first expansion of German speakers into the upland basin was prompted mainly by the need for expanded pasturage. We already have seen how the monastery at Unser Frau and the initially German, though later Romance, village of Ruffré, below the Mendel Pass, played a part in regulating the movement of livestock from villages in the Etsch River valley to summer pastures on the Anaunian slopes of the Mendel ridge. Proveis, too, first served as a pastoral station for herds from the valley of Ulten.

The need for additional pasturage at this time resulted from population growth at lower altitudes. The growing settlements in the valleys forced the conversion of pastures lying at low or middle altitudes to meadow; hence, grazing lands had to be found ever farther up the mountainside. Selection of the Anaunia for new pasture ground was a logical choice for villages located between Lana and Kaltern along the western wall of the Etsch River valley: the slopes on the Etsch side are precipitous, whereas those falling off toward the Noce are gradual and, therefore, better suited to grazing.

The movement of German speakers into the Anaunia was not an isolated event, but a part of the general movement toward colonization of the high valleys in the Tyrol, which was a response to the forces of commercialization and political centralization. Furthermore, the conversion of temporary herdsmen's camps to permanent, year-round settlements by cultivators was not merely an answer to local population growth and the need to feed an additional number of mouths. It also provided an outlet for Tyrolese displaced by the influx of foreign labor into the mining industry, an influx that simultaneously increased the demand for land and closed off opportunities for non-agricultural employment. The emergence of the settlement around Unser Frau thus was followed, in the fifteenth and sixteenth centuries, by the expansion of the area settled by German speakers down-valley to Malgasott and St. Felix (then Caseid), and the subsequent intensification of settlement throughout the Upper Anaunia. Yet the growth of permanent settlements such as St. Felix and Tret inevitably reopened the cycle of conversion of pasture to home fields and meadows that sent village livestock still higher up the mountainside in search of grazing land.

At the same time, the possibilities for establishing new home fields and villages in the Upper Anaunia were limited by factors of altitude and slope. Even today only slightly more than 11% of the land surface is suitable for cultivation; very little of the land lies below 900 meters, the upper limit for the cultivation of commercial orchards, and only a few holdings located at the lower limits of Castelfondo, Laurein, and Rumo can support apple orchards with any degree of success. Moreover, much of the land is simply too steep to use at all. The home fields are almost all found at 900–1300 meters, although a few plowed fields do reach the 1400-meter mark and meadows are occasionally pushed up to 1500 meters. However, even at 1300 meters, climatic conditions are sufficiently severe that peasants with home fields above that contour complain that they must put more land under the plow and use more seed per hectare to obtain harvests comparable to those produced but 100 meters downhill. In one place, at 1500 meters *Auf der Klamm* (on top of the canyon) in St. Felix, there are traces of fields once plowed and now abandoned—evidence of an attempt in years gone by to establish an estate above today's limits. Since the attempt to use these fields was made only once, despite increasing hunger for land, it is clear

"A few plowed fields do reach the 1400-meter mark and meadows are occasionally pushed up to 1500 meters." Upper limits of settlement, Auf der Klamm, St. Felix.

that cultivation in this location had reached its upper ecological limit.

Not only did expansion reach an environmental limit at St. Felix; it also encountered political limits. There is no specific documentary evidence of resistance to the grazing of German cows on Romance pastures in the early Middle Ages. Yet later expansion, and especially the conversion of forested mountainside above St. Felix and Tret to mountain pastures and meadows, was fraught with contention, not only between Romance and German communities, but among German communities as well. That disputes over rights of access to mountainside, pasture, and forest should have persisted a long while is not difficult to understand when one examines the present-day interspersion of parcels of ground owned by people from adjacent villages. Although the deep gorges of the tributaries to the Novella—the Mühlbach and the stream rushing down the Klamm —provide convenient and easily agreed upon lines of demarcation between Unser Frau and St. Felix, and St. Felix and Tret at lower altitudes, such natural boundaries do not exist farther up the mountain. Parcels of mountainside were appropriated by individuals for meadow, timber land, or pasture in quite irregular fashion, so that village boundaries could not be drawn without including fields held by members of rival neighboring communities. Thus, villages inevitably came into conflict over claims to village common.

Intervillage strife persisted, moreover, as each community sought to include within its own political confines all the land used by its members.

As the area grew in population, boundary lines became both ecologically and politically relevant, and they were ultimately put under the supervision of the territorial state. Consequently, further expansion down-valley or into adjacent valleys by German speakers from Unser Frau and St. Felix was met by opposition from German-speaking Prissian and Tisens, as well as from Romance-speaking Rumo, Castelfondo, and Fondo. Tret, however, as a ward of Fondo, could expect the council of the larger commune to contain St. Felix politically within its natural ecological limits.

By the end of the sixteenth century, therefore, the expansion of home fields in Unser Frau and St. Felix came to an end. Not only are the names of holdings extant in the sixteenth century still used, but no new ones have been added. Moreover, the twenty-three *Hofrechte*, or codes of pasture rights, in force today correspond exactly to the number of holdings in existence around 1600. Subsequent expansion in the number of estates has taken place by a division of existing holdings, rather than by the clearing of new land: the attempt to create a new estate above 1500 meters at *Auf der Klamm* in St. Felix failed.

The settlement area occupied by the twenty-three homesteads at the end of the frontier period, 1600, measured about 150 hectares of land, which could be devoted to meadow and plowland, or an average of just over 6.5 hectares of homeland available to each holding (see Figure 10). In fact, reconstruction of the layout and size of the original estates in St. Felix shows that all except one contained 2–10 hectares, the exception rising above this range to 15 hectares. Mean estate size is slightly above 6 hectares. This figure is an approximation of the amount of land needed to provide food for one household—excluding workers not members of the domestic unit—under conditions of subsistence agriculture.

With nearly all land under plow by 1600, further expansion was not feasible. The meadows and lands under cultivation today exceed by just slightly over 10 hectares the total calculated for the original homesteads. Although the attempt to push beyond the 1500-meter contour failed, there are several holdings in St. Felix today which, though located at lower altitudes, occupy very poor soil. These number among the poorest households in the village, and they must bring in outside earnings to survive. Yet even in 1961 such earnings did not guarantee an adequate livelihood: access to land is essential for survival in the community. St. Felix seems to have reached not only its ecological and political limits, but also the ceiling on demographic expansion.

The pressures upon the population of Romance-speaking Tret are similar, but not identical. In the German-speaking villages, communal councils have regulated internal affairs and defended village interests against outsiders at least since the middle of the fourteenth century. Tret, however, has been a political dependency of Fondo as long as written records have been kept; the commune of Fondo as a *whole*, rather than the *frazione* of Tret, has jurisdiction over the communal lands surrounding the home fields of the village. Thus, the decision to convert pasture and forest to meadow and plowland in Tret responded not merely to the requirements of the Trettners, but also to wider interests, most especially those of landholders in Fondo who wanted to maximize their access to grazing lands and forest. While the German-speaking communities bordering upon Tret expanded their home fields to the ecological limit, these limits have never been reached in Tret. Even in the twentieth century there is still some land that could be cleared for new homesteads. In general,

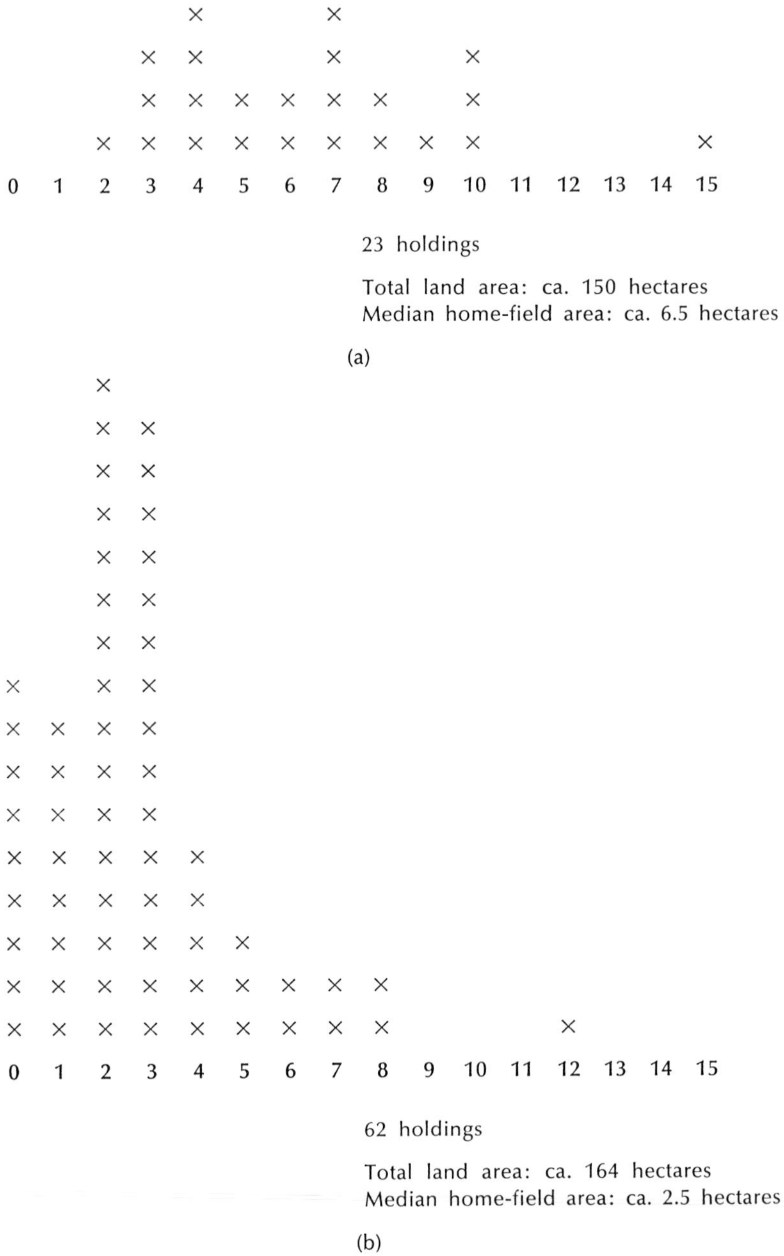

```
              ×           ×
          ×   ×           ×               ×
          ×   ×   ×   ×   ×   ×           ×
      ×   ×   ×   ×   ×   ×   ×   ×   ×                       ×
  0   1   2   3   4   5   6   7   8   9   10  11  12  13  14  15
```

23 holdings

Total land area: ca. 150 hectares
Median home-field area: ca. 6.5 hectares

(a)

```
          ×
          ×   ×
          ×   ×
          ×   ×
          ×   ×
          ×   ×
          ×   ×
      ×   ×   ×
  ×   ×   ×   ×
  ×   ×   ×   ×
  ×   ×   ×   ×
  ×   ×   ×   ×   ×
  ×   ×   ×   ×   ×
  ×   ×   ×   ×   ×   ×
  ×   ×   ×   ×   ×   ×   ×   ×   ×
  ×   ×   ×   ×   ×   ×   ×   ×   ×               ×
  0   1   2   3   4   5   6   7   8   9   10  11  12  13  14  15
```

62 holdings

Total land area: ca. 164 hectares
Median home-field area: ca. 2.5 hectares

(b)

Figure 10. Home-field hectares, St. Felix: (a) original holdings; (b) current holdings.

the council of the commune of Fondo has resisted petitions from would-be homesteads, but has occasionally yielded to individual demands for homesteads by selling parcels of commons; on at least two occasions—1825 (Inama 1931: 45, fn. 1) and 1930—the council has permitted major sales. Restrictions on land use in Tret are then primarily political rather than environmental.

THE FACTS OF LIFE

Economic and political checks on the expansion of arable and pasture land, however, posed a serious problem to the mountain communities: economic resources were subject to limitation, but villagers could not similarly be limited in the number of offspring they produced. Generation after generation, St. Felix and Tret have harvested a crop of children far larger than either village could support to maturity. One solution has been to divide existing holdings: by 1800 there were nearly twice as many estates in St. Felix as two centuries earlier. Yet division causes its own problems: holdings of insufficient size and not equipped with the necessary range of resources will fail or be reabsorbed into viable estates.

A second response—intensification of production on a given area of land—was possible after 1800, when the potato was introduced. Cultivation of the potato resulted in a marked increase in total productivity in both St. Felix and Tret, as well as a renewed growth of population. Following 1800 the rate of property division accelerated, so that by 1879 the number of homesteads in the villages was about three times the number in 1600, when the Anaunian frontier closed. Yet toward the last quarter of the nineteenth century, the population crisis returned.

The villagers found that while the potato indeed permitted a more intensive exploitation of the land, more food raised on the same area merely increased the number of people dependent on the same limited supply of resources. The surrounding lowlands were moving toward the production of cash crops, expanding their commercial involvement, widening the area under cultivation through drainage of marshland and swamps, and using the newly constructed roads and railroads to transport produce to new and untapped markets. In contrast, mountaineers on the high reaches of the Anaunia only rarely produced surplus crops, nor could they hope to raise potatoes in sufficient quantity to compensate for the low unit price and high transportation costs created by the inaccessibility of their villages. Furthermore, employment outside agriculture remained limited and temporary; as there were many hungry mountaineers eager to work outside their villages, wages barely sufficed to maintain a worker during his short periods of employment. Even today many villagers remember pre-World War II times, when earnings covered no more than room and board, leaving a few coppers for wine on Sunday.

The terms of this fundamental Malthusian equation, in turn, created a set of contradictions that every villager had to confront in the course of his life. Land remained the basis of livelihood, and production for subsistence continued to be the goal of every manager and owner of land. With little chance for employment outside the village, access to land alone assured a measure of subsistence. On a limited land base, therefore, each successive generation of villagers was subjected to the same fierce selection pressures that separated the holders of viable homesteads from those of estates no longer viable, and pitted villagers with access to land through inheritance or purchase against villagers denied such access.

Thus, the differential distribution of rights to own or manage land was the crucial issue in village life until the middle of the twentieth century. We shall examine in some detail how Felixers and Trettners have coped with this problem. Only with the postwar growth of the European economy have there appeared real solutions to this grim struggle for survival. We witnessed, during our fieldwork, the advent of alternative careers for village youths and increased market opportunities that have begun to change the conditions under which the Anaunia has labored for more than seven hundred years.

CHAPTER **VII**

The Mountain Estate

As we saw in Chapter VI, the number of estates in the Upper Anaunia has grown over the years and average estate size decreased as a consequence. The composition of estates has changed from time to time as a result of property divisions by bequeathal and through the sale of land parcels. Yet the processes of expansion and contraction have been worked out for the most part step by step over extended periods of time.

When a villager takes over management of an estate, he does not anticipate any grand change in the composition of the holding nor in his control over other resources: he expects the estate to remain relatively unaltered throughout his entire life. Perhaps the estate will even remain intact after his death, to be managed by his heir. Or, perhaps there will be a rearrangement of his holding and that of his neighbors, as the result of property division caused by marriage or migration.

However, the disposition of property has no direct effect on the way the manager himself operates the estate. He is concerned with the daily routine and with a land-use strategy that will make the year successful. Since his resources change little from one year to the next, and since the annual work cycle is essentially unvarying, long-term planning is not complex: it consists in large measure of making a success of one year at a time. While the estate manager may hope to obtain a new field or meadow, or to increase his small herd of cattle by raising one more calf to maturity, his calculations do not normally include reinvestment. The economics of subsistence farming do not include the concept of growth. If there is sufficient fodder to see the animals through the winter, if house and equipment are in repair, and if there is a bit of money hidden away somewhere in the house, then the year has been good indeed.

Estates differ in size and productivity; consequently each estate has a fixed potential. Villagers are fully aware of the productive potential of village estates, and they have a good idea of how closely potentials are being realized. The measure of a man as a proper peasant is based upon how well he runs his estate. Thus, although the amount of land available to any producer is a constant, production may vary in accordance with labor expended. To maintain a successful estate, the manager and his household must devote sufficient energy to the estate, under skillful direction.

RIGHTS IN ESTATES

The right to benefit from the production of an estate depends on membership in the domestic unit that holds the estate or on participation in its operation. Generally one belongs to a domestic group through birth or marriage, but adoption is practiced in the Upper Anaunia, and unrelated individuals are sometimes 'hired' to work on the estate for a finite period of time, for the duration of which they are incorporated into the household routine.

The central figure in each domestic group is the estate manager. In most cases, he has come to this position through an inheritance process that places the estate under the former manager's heir, thereby leaving certain kinsmen dependents of the new manager (see Chapter VIII). It is expected that the manager of an estate will also be its owner; that is, the individual who decides how the estate is to be operated and how its products are to be distributed will also be in a position to sell, rent, or mortgage the estate. The term "ownership" implies private property—hence, "ownership" can be used here only with caution. In the operation of the estate, but especially in the disposition of its produce, and in the sale of the holding, the manager is constrained by other rights and claims to the estate. These constraints have their origin in external relations of the estate and in its internal social and property relations.

We have already seen how managerial rights to estates evolved from feudal obligations to free hold over the land. Property deeds to land are now made out and recorded by the state; the deeds registry office for all the Upper Anaunia villages was located in Fondo until the mid-1960s, when records for the German villages were transferred to Meran. The registered owner(s) has the right to transfer ownership through inheritance or sale, or to mortgage the land. However, since the late Middle Ages at least, the village communities at times have interposed themselves between estate holders and the market, exercising the right to approve or disprove sales agreements and wills. We noted that the Fondo council used such rights to prevent the sale of local estates to outsiders or to members of the nobility. In more recent times, however, regulation of property sale has been more common in areas of German speech. The laws passed by the Tyrolese assembly to prevent the division of estates in the German Tyrol were enforced at the village level by community councils made up of locally elected villagers. Yet the laws did not reach the Upper Anaunia as these villages were included in the Trentino until 1948, when they were transferred to the South Tyrol. None of the holdings was officially classified as "closed" until after the transfer, but village sympathies favored these laws and property division was resisted by village opinion. This position is well illustrated by the voluntary designation of fourteen holdings in Unser Frau and one in St. Felix as "closed," as permitted by the estate laws established in the South Tyrol in 1954.

Since that time each community has elected a local estate commission (Örtliche Höfekommission) that regulates all permanent transfers of land ownership. While the operational instructions for the commission are handed down by the provincial council (Landesrat), the membership is elected locally. Composed of respected, local men well acquainted with village affairs and backed by the state, the commission is able to enforce its decisions. It zealously refuses to permit the break up of holdings either through division by inheritance or piecemeal sale of parcels. The commission will permit the detachment of parcels from a holding only if they have a history of sale or have been brought into the homestead as a dowry. Not only is the division of holdings classified as impartible prohibited, but all land sales require the commission's approval—and it looks askance at the detachment of land from any holding, whether classified impartible or not. The commission may also prevent the sale of land within the village to outsiders, especially to Italians.

In one instance, a villager in St. Felix sold several parcels from his estate to a Trettner. The man who sold the land was an alcoholic, and the land had not been worked as an integral estate since the death of his mother. Part of it was being worked by a villager appointed by the court as the owner's guardian and who had received the right to work a portion of the estate in compensation. The remaining land was rented to several villagers. When the owner decided to sell, the village council was opposed to division of the estate, even though the owner was not able to work the land. The Council was able to take legal action on the grounds that the owner, as an alcoholic, was not legally responsible, and the sale was therefore invalid. This move was successful, and the land was returned to its owner. The purchaser, incidentally, was never reimbursed for the money he had paid for the land. All he could do was to shrug this off as one of the minor misfortunes that are an inevitable part of life.

While community regulation is exercised only in cases of the alienation of land from an estate, the manager may carry obligations to a number of categories of individuals who can regularly make demands on the output from his estate. These individual claims largely have their basis in domestic relations, but they are sometimes sanctified in the jural realm as well. Succession to management of the estate involves not only rights and privileges, but also obligations. Members of the domestic group, especially siblings and parents of the manager, retain rights to a share of the estate's produce: the new manager can expect claims to be made by his parents, by siblings who, for whatever reason, have not declared themselves free of the estate, as well as from his spouse and children. The basic assumption is that any individual born into a domestic group retains the right to a share of the produce of the estate as long as he remains a member of that group. Such membership is demonstrated by ongoing participation in the affairs of the estate.

Rights to a living from the land are held by the retiring manager and his wife for the duration of their lives. Villagers agree that it is the obligation of the new generation to care for parents, but the burden of support falls to the new estate manager. A former estate owner's claim to usufruct is strengthened if he chooses to retain ownership of the land until his death, instead of relinquishing it when he retires from active management. Even when ownership is transferred to the heir, jural rights to a share in the fruits of the estate will be protected by assigning the former owner legal usufruct (German: *Fruchtgenuss*, Italian: *usufrutto*). While this will be written into the deed as the right to a specific percentage of the produce (perhaps one-quarter), in practice the provision means that he will receive a room and bed, clothing, and a place at the table of the new manager.

The rights of a widow to a living from her husband's estate is more complex. A widow's rights to the estate are derived in part from her marriage to the manager, and in part from her status as mother of his heir. Ownership of the estate passes to the legitimate offspring of a man at his death. His widow does not share in the inheritance, but is assigned the right to a share of the estate's produce. While village sentiment places the responsibility for her support on her offspring, especially the estate manager, her position is also protected under Italian law, and her rights as a widow will be entered in the property deed. Any land she has brought with her into the marriage, or that has been purchased in her name, remains her own. It is never transferred to her husband, and she is legally entitled to its produce, although in practice it will continue to be operated by her husband's heir as an integral part of the estate. It will, however, be disposed by her will.

If her husband dies without an heir, the widow's position is much less secure, especially if she is still young. Ownership of the estate will pass to the owner's siblings, and although his widow is legally entitled to a living from the land, it is unrealistic for her to expect to be supported. The estate will be used to support the new owner, and there is no social provision in the Upper Anaunia for the inclusion of a young widow into such a household—the only way an individual has access to the produce from an estate in this subsistence economy.

While usufruct is most often demanded by the parents of an estate manager, such rights can also be assigned to other individuals. This is most common in the German villages, where it is used to provide for unmarried daughters. Since ownership of the estate in these villages is preferably assigned to a single heir, his siblings have no *legal* claim to further support from the estate, and their *personal* relationship to the new owner thereafter determines their success in claiming a share in its affairs. Parents are reluctant to interfere with the heir's control over the estate by assigning other siblings part ownership. Concerned with the well-being of all their offspring, however, they sometimes provide usufruct for a daughter in their will; she will have rights to a room in the house and to a share in the estate's produce until she marries.

The estate manager is always faced with actual or potential claims on the estate from his siblings, whether such claims are based on legally recognized usufruct or not. In Chapter VIII we will describe the process whereby each generation is sorted into those who inherit and those who do not, and the ways in which social relations within the domestic group develop around access to property. Here it will be useful to outline the types of relationships that result from this process. In the German-speaking Tyrolese villages a single estate manager is the ideal. While siblings share equal rights to support from the land as long as their parents are alive, they no longer hold this right when the parents die and are succeeded by a chosen heir. Concern for the disinherited offspring may take the form of a provision that the heir make a cash settlement on each of his siblings. However, this provision does not carry a moral or legal claim on the holding. The siblings may, however, retain a claim to support from the land by continuing to participate in its operation, either as full-time resident hands, or by occasional gifts of cash and donations of labor at crucial times of the year— such as at the harvest. However, the principal heir is free to accept or reject such participation. Whether a manager accepts full-time or fringe participation in the estate's operation by his siblings is determined by self-interested strategies.

Tyrolese estates have single manager-owners; Nónes estates, however, are owned more frequently by sibling sets or subsets. Here the ideal is that wills provide each offspring with an equal share in the ownership of the estate. This ideal is not always realized, but several offspring usually do share ownership of the holding. As we shall see, this group will in time sort itself out, as some individuals become managers and others leave the estate. However, as long as an individual holds a share of the estate, he retains the right to participate in its operation and to benefit from the harvest. While the sorting process generally leaves one individual in managerial control, he can expect at least some of his siblings to participate in the estate's affairs, in either a core or a fringe relationship. Yet he has less control over the situation than his German-speaking counterpart, since the claim to a share of ownership held by any sibling will ensure his rights to support should he choose to contribute to work on the estate. Ownership without participation, however, cannot command a share in the estate's produce: the individual who is uninterested in contributing to the estate's operation can expect no more than a cash settlement. He cannot collect a rent because he owns part of the land. This makes sense in terms of the subsistence orientation of the Upper Anaunian economy.

Still, the manager of the estate is constrained in his operation of the holding even by those who leave the estate as long as they retain part ownership thereof. This becomes an acute problem if the manager wishes to treat his land as capital by mortgage or sale, or to transmit ownership to his heirs. Here he is constrained by the law, which recognizes the absent part owners' rights. Market operations cannot be conducted without the participation of all those listed as owners on a deed, whether or not they participate in the affairs of the estate. This problem is compounded in cases that have not been resolved for more than a generation: the manager may have control of an estate and its resources, but be unable to sell or mortgage the property because of the outstanding share. After a generation or two, communications may break down with the former part owner and his heirs, so that the manager may know neither the identity of the part owners nor their whereabouts. Every manager in this situation lives in terror of a distant cousin suddenly appearing, to demand a division of the estate on the basis of some ancient claim.

In one illuminating case, a one-third share of a medium-sized holding in Tret passed through three generations without its ownership being settled on the manager. The part owner had migrated to Argentina before 1920. On his death he left eight heirs, and by 1960 some of these had died as well, leaving additional heirs. By 1960 there were twenty individuals living in Argentina, all with some measure of claim to the ancestor's one-third share of the estate. The Argentine branch of the family had prospered, first as cattle ranchers, then in various professions. When one of the numerous heirs, a lawyer, decided to visit the village while carrying on advanced studies in Italy, the manager of the estate was extremely apprehensive, fearing that the visitor might be returning to take up life on the mountainside, and to claim his grandfather's share of the estate. These fears may seem unreasonable, but they were very real to the manager. Even though the visiting lawyer proved helpful in assisting the manager to gain complete ownership of the holding, the manager's apprehensions were still evident five years later: recalling those traumatic times during an interview resulted in severe asthma attacks on two occasions.

ECONOMIC PURSUITS

Difficult as was the life of a mountain peasant, the alternatives available to the sons of the Upper Anaunia before 1950 offered poor returns in comparison. Without at least some claim to a holding, one could not earn a living within the village; at the same time, careers outside the area that offered continuity and a reasonable income were rare. Most employment available in the surrounding region was short term, varying from a few months to one or two years, so that the careers of the landless involved continual movement from one job to another; periods when no work at all could be found were frequent. Yet, the Upper Anaunia has always produced a population much larger than the local economy could support, forcing a sizable percentage of each generation to seek alternatives to the role of landowning peasant. Nearly 40% of those born since 1800 and who have survived into adulthood left the village for good. Another 24% remained single throughout their lives, seeking careers outside of the village, yet never were able to become completely independent of their natal holding (see Tables 4 and 5).

Under such conditions, competition for land has always been keen. Individual strategies have been directed toward obtaining the best possible claim to as much land as possible. To succeed in these operations, the most important factors were the size of the holding and the strength of one's claim to the land. At any point of time in the past, one can think of the adult inhabitants of the villages as arranged on a continuum according to the ability with which they supported themselves within the local community. At one extreme of this continuum stood the largest landholders, men able to support families exclusively from their land. At the opposite extreme were those who had either no land or held rights to so little land that remaining in the village was out of the question, and who therefore permanently emigrated. Useful reference points can be marked off along the continuum to describe economic situations of villagers in the past and the present (see Table 6).

In order for a domestic unit to be fully supported by a holding, the estate not only had to be relatively large, but also encompass a balanced combination of varieties of land. There had to be enough plowland to raise a range of crops in sufficient quantity to feed the family for the entire year and to provide some of the fodder for its animals. Sufficient meadow land had to be held, both within the village and on the mountainside, in order to produce enough hay to feed perhaps a dozen cattle, one or two draft animals, and eight to twenty sheep throughout the long winter. In addition, the family required access to forestland, through private ownership or through rights to communal land, where livestock could graze and lumber and firewood be obtained. To fall within this category a holding needed over six hectares allocated to plowland and village meadows, with an additional ten or more hectares held in mountain meadows and forest (in Tret, as indicated above, rights to the communal forest were substituted for privately owned woods). With the produce from such a holding, a family could meet virtually all its needs and expectations, and raise, mainly through the sale of cattle and lumber, whatever cash was needed. To do so, however, demanded the full energies of the family, and often on this size holding, the family labor pool was not sufficient. It could be supplemented in one or more ways: by enabling a sibling of the owner or his spouse to reside on the holding, by keeping a full-time "hired man," or by hiring day labor at periods when it was most critically needed, as at the hay harvest.

The number of such relatively large holdings has never been great in the Upper Anaunia, averaging around 10% in this century. Roughly half of the holdings were smaller, falling into a category of middle- and small-sized holdings. These possessed the full range of varieties of land in sufficient quantity to supply a minimal living. Yet a family with such a holding could not, by agriculture and livestock-keeping alone, supply adequate material and cash to meet culturally defined expectations, or outside pressures for money, without supplementing its income. Both lime "burned" out of local stone and charcoal produced from the wood of private or communal forests could be sold in neighboring markets. Men from both St. Felix and Tret with talents for carpentry made wagons, plows, and sleds for sale to neighbors; others worked as coopers, millers, or roofers, and several villagers operated sawmills. In the German-speaking village, the making of wooden hay rakes and scythe handles was a popular craft; the finished product was carried over the Gampen Pass by the peasant and his son, to be sold at some village market or peddled from holding to holding in the Etsch River valley. While these sidelines were important in supplementing "in kind" and in cash the earnings from the peasant's holding, most of the working hours of a householder were devoted to his land and livestock. By and large the labor requirements of the estate could be met without outside assistance. Whenever additional labor was needed, it was usually

| Decade born | Population surviving past age 20 | | | Married, lived locally | | | | | |
| | | | | Total | | Natal holding | | Other[a] | |
	Total	M	F	M	F	M	F	M	F
1920–29	60	26	34	20	23	11	6	5	13
1910–19	39	15	24	10	13	5	4	2	8
1900–09	47	31	16	17	7	6	0	5	7
1890–99	47	18	29	11	22	3	5	3	8
1880–89	89	51	38	24	15	9	0	4	9
1870–79	60	31	29	21	16	7	0	4	10
1860–69	62	38	26	19	13	7	1	6	4
1850–59	107	50	57	37	30	11	1	7	16
1840–49	68	25	43	21	24	8	2	6	16
1830–39	42	22	20	15	8	4	1	7	3
1820–29	61	32	29	28	18	8	3	5	7
1810–19	57	30	27	19	19	6	2	3	6
1800–09	56	31	25	19	12	8	4	1	4
Total:	796	400	396	261	220	93	29	58	111

[a] Married into holding in either St. Felix or Unser Frau.

[b] Includes both full- and part-time residents.

[c] Since contact with the village was not maintained with all of these individuals, some of them may eventually have married or joined the church, and many are thought to have migrated to America.

| Decade born | Population surviving past age 20 | | | Lived in | | | |
| | | | | Total | | Married | |
	Total	M	F	M	F	M	F
1920–29	46	28	18	16	4	9	3
1910–19	29	16	13	5	4	5	3
1900–09	74	36	38	15	11	11	10
1890–99	55	25	30	11	9	9	4
1880–89	81	36	45	16	12	11	8
1870–79	79	37	42	22	14	15	6
1860–69	56	27	29	10	13	8	9
1850–59	56	37	19	21	7	12	5
1840–49	44	24	20	16	8	13	6
1830–39	33	16	17	9	3	7	1
1820–29	68	36	32	15	4	12	2
1810–19	48	26	22	6	3	6	1
1800–09	31	14	17	5	2	5	2
Total:	700	358	342	167	94	123	60

[a] Includes both full- and part-time residents.

[b] Contact with the village was not maintained with all of these individuals. Some of them may have married or migrated. It also includes those who joined the church.

ADULTS IN ST. FELIX

Single, lived locally[b]		Left village permanently									
		Total		Married		Single[c]		Migrant[d]		Church[e]	
M	F	M	F	M	F	M	F	M	F	M	F
4	4	6	11	1	5	3	6	0	0	2	—
3	1	5	12	0	2	4	7	1	2	—	1
6	0	14	9	5	8	8	0	1	1	—	—
5	9	7	7	2	3	3	4	1	0	1	—
11	6	27	23	4	8	6	8	10	0	7	7
10	6	10	13	3	6	2	2	3	0	2	5
6	8	19	13	6	7	7	2	3	0	3	4
19	13	13	27	3	14	8	9	2	0	—	4
7	6	4	19	1	7	3	11	—	—	—	1
4	4	7	10	2	2	5	8	—	—		
15	8	4	11	1	2	3	9	—	—		
10	11	11	8	2	0	9	8	—	—		
10	4	12	13	0	0	12	13	—	—		
110	80	139	176	30	64	73	87	21	3	15	22

[d] These are permanent migrants, mostly to the Americas. Other villagers who went there and returned are included as village residents.

[e] This includes men who became priests or monks and women who became nuns or housekeepers for priests.

ADULTS IN TRET

Tret Single[a]		Left village permanently							
		Total		Married		Single[b]		Migrant[c]	
M	F	M	F	M	F	M	F	M	F
7	1	12	14	5	9	6	4	1	1
0	1	11	9	3	8	7	1	1	0
4	1	21	27	1	13	14	13	6	1
2	5	14	21	0	14	10	7	4	0
5	4	20	33	4	22	13	7	3	4
7	8	15	28	1	14	7	11	7	3
2	4	17	16	1	10	4	6	12	0
9	2	16	12	2	9	11	3	3	0
3	2	8	12	0	5	8	7	0	0
2	2	7	14	1	4	6	10	0	0
3	2	21	28	0	8	21	20	0	0
0	2	20	19	1	6	19	13	0	0
0	0	9	15	0	2	9	13	0	0
44	34	191	248	19	124	135	115	37	9

[c] These are permanent migrants, mostly to the Americas. Other villagers who migrated and returned are included as village residents.

Table 6

CURRENT HOLDING SIZE[a]

Size (hectares)	St. Felix (excluding forest land): No. of holdings (%)		Tret (excluding forest land): No. of holdings (%)		St. Felix (including forest land): No. of holdings (%)	
Dwarf: up to 3	16	(25.8)	21	(42.0)	9	(14.3)
Small: 3–5	15	(24.2)	7	(14.0)	6	(9.7)
Medium: 6–10	21	(33.9)	16	(32.0)	19	(30.7)
Large: 11–15	7	(11.3)	3	(6.0)	14	(22.6)
Large: over 15	3	(4.8)	3	(6.0)	14	(22.6)
Total:	62	(100.0)	50	(100.0)	62	(99.9)
Mean holding size (hectares):	4.5–5.0		4.0–4.5		10	

[a] Current holding size in this table represents the total amount of plowland, meadow, and mountain meadow held by each estate. Forest land in Tret is owned communally, each estate holding rights to an annual allotment of timber. Estates in St. Felix own their own timberland.

obtained by labor exchange, established either on a contractual basis with a neighbor, or through kinship ties to close affinal or consanguineous kinsmen.

There were in St. Felix and Tret various specialists who worked middle-sized or slightly smaller landholdings, but who also devoted a major portion of their working hours to specialties. The blacksmith and the more popular of the carpenters were kept busy throughout the year at their trades, somewhat neglecting the land, and leaving much of the agricultural work to their already overburdened wives. Innkeepers and store owners, however, left most of the shop-tending to their wives; they were peasants first and merchants second.

Forty percent of the landowners were men with little land (dwarf holdings), but with a house and enough land to keep one or two cows, or a few goats. They not only held insufficient land to provide full support for their families, but often lacked an even balance in varieties of land. Such individuals were not able to remain full-time inhabitants of the village year in and year out, but were forced to seek seasonal employment in the surrounding lowlands. Those with the most meager holdings had to seek additional employment almost continuously. Such individuals were seen in the villages only at planting and harvest time—unless they were lucky enough to find frequent work as a day laborer for one of the larger land-

Blacksmith shoeing a horse, Tret.

holders within the village. In Tret, stonemasonry was a craft specialty practiced by many of the small landholders. Most of their work was found in the German-speaking valleys over the Gampen Pass to the north, where the population has traditionally relegated such work to Italian specialists. Even in St. Felix and other small Tyrolese peasant villages, the building and repair of stone buildings is done by hired labor. This work was seasonal, but enough was available so that the Tret mason—a small landholder as well—could anticipate a total annual income that approximated his culturally motivated expectations.

Still other villagers either had no land or but a field or two, or (mainly in Tret) their ownership rights to a larger holding were insufficient to provide a living from the land (for example, a man might inherit one-sixth of an undivided, medium-sized holding, while his brother or cousin accumulated five-sixths ownership). Many of them left the villages. Although emigrants from the Anaunia have scattered to all continents, in this century the largest movement has been to the Americas. Villagers are fond of pointing out that there are more people born in Tret and St. Felix but now living in the Americas than remain in the villages. Although there were a few migrants to the Americas before 1900, the overabundance of labor in Italy and Austria around the turn of the century, coupled with an economic crisis, caused a sudden and dramatic wave of migration to the New World (Clough 1964: 124–140). Most of the migrants never returned. Likewise, many of the landless left the villages to enter the Church. Virtually every family in St. Felix and Tret can point to uncles who became priests or monks and to aunts who became nuns or housekeepers for their priestly brothers.

A small minority of landless villagers were supported on the holdings of a landed sibling, receiving payment in kind for work done on the estate. Some of these were able to obtain a degree of independence by developing an economic sideline—operating the family mill, hauling lumber for the sawmill or for peasants with lumber to saw, working as day laborers for other peasants, or trading in livestock. This category shades into another: men and women who left the village to find work in the lowlands, but who retained ties with the villages and the holdings where they were born through visits and occasional periods of residence there. These disinherited siblings still played upon kinship sentiments to ensure periodic support from the land when work could not be found elsewhere, and they kept such ties functional by timely donations of labor and contributions of cash. Although opportunities for full-time employment outside the village, especially employment that could be expected to last through the years, were limited, some individuals managed to make their way through life as herdsmen, farm laborers, teamsters, animal traders, lumbermen, and stonemasons (Trettners)—a rare individual became a beggar. Women sometimes found employment as domestic servants in the houses of wealthy lowland farmers or in city homes, as waitresses, and as chambermaids in hotels.

Where the sideline or career was entrepreneurial and money was manipulated, additional hazards were involved. To establish a teamster service, to engage in animal trading, or to operate a sawmill one has to extend credit and borrow money from outside sources. Given the shortage of money in this subsistence economy, it is not surprising to learn that entrepreneurs are eventually faced with increasingly strong demands to pay off debts, whereas peasant-customers, lacking cash, delay payment for services and goods received. The more timid generally would give up their sideline at this point and pay off their debts from income as wage earners and the occasional collection of payments long overdue from former customers. Many, however, delayed capitulation by exploiting kinship ties, by persuading landed kinsmen to mortgage their land against mounting debts in return for a share in future profits. Such tactics only delayed the inevitable collapse, and resulted in disaster for all involved. When finally faced with a court order to pay, with no way to collect due money, and with no new sources of capital available, the entrepreneur had no choice but to yield, forcing his backers into bankruptcy and loss of their holdings.

The peasant owner-operator of a holding usually acquired his estate through inheritance, and land changed hands through sale only as the result of a crisis such as bankruptcy or the dying out of a family line. Despite the rarity of sale of an entire holding, individual plots of land were occasionally sold. Every landholder, except possibly a man with as much land as his household could work, was constantly on the alert to increase the size of his holding, and men without land worked and saved for the day when such an opportunity might come their way. Frugal sons who did not share in the inheritance of their father's lands sometimes saved for a lifetime without finding a suitable holding to buy. The goal of many migrants to the Americas was to earn enough to enable them to return and buy a holding of their own—some were successful, but many returned to the villages to buy land, spent their fortunes supporting themselves instead, and departed again without having acquired the hoped-for estate.

While chances to buy a holding outright were rare, an individual with a small bankroll could sometimes marry the daughter of a landowner who had no sons. Such a son-in-law was often preferred by the girl's father to one with a holding of his own, especially if the father was in debt. While the bridegroom would not become owner of the holding and would have a low status compared to his landowning neighbors, he would obtain a measure of security in his lifetime not ordinarily available to landless men, and he could take solace in the expectation that his sons would inherit the estate.

Except in the face of an extreme crisis, when a large amount of money was needed at one time, the large and medium landholders were permanent residents of the village, representing the core of the village population. In the most stable material position, these men were also accorded the bulk of the social and political honors. The men selected as political leaders always came from this group, and the principal political figures in each village, the *Bürgermeister* in St. Felix and the *capo frazione* in Tret, were almost without exception peasants considered the wealthiest men in their villages. With sufficient land, a man could marry and raise a family. Without land, marriage was virtually an impossibility: by far the majority of the landless remained single throughout their lives, and those who did marry led lives in which long separations from their families were the rule. The separation often lasted for years on end, while the man labored abroad trying simultaneously to support himself and his family, and to save enough money to buy a holding of his own.

THE ORGANIZATION OF LABOR

Most of the labor on an estate is provided by the members of its domestic group, but there are tasks that, either by necessity or choice, require supplemental labor. With rare exceptions, this labor is recruited within the Upper Anaunia, but the method of recruitment and the nature of labor relations vary both with the size of the estate and in response to the differences in social organization between Nónes and Tyrolese villages.

In St. Felix, estate managers strive for complete self-sufficiency. Even ties to the manager's siblings are retained only when deemed economically valuable, and cooperative work relations established with in-laws during the courting period are rarely of long duration. When it is necessary for the manager to obtain help from outside the domestic unit, strict accounting is maintained. Close track is kept of the amount of labor used, and each party is careful to make sure that exchanges are equal. Ideally, exchanges cancel each other out exactly, so that no one leaves the transaction beholden to anyone else. Estate managers rarely exchange visits except for the purpose of conducting business, and then only when the matter cannot wait for Sunday to be discussed at the village inn.

In Tret, however, exact count of the amount of labor and working time spent is rarely calculated, unless labor is to be paid for in cash. Visiting back and forth between apartments is carried out by both related and unrelated domestic groups. Several special sorts of services are provided within the village by individuals who receive no direct recompense for their work: two brothers with a horse-drawn plow keep village by-ways free of snow, and a villager who owns a small pistol slaughters pigs owned by fellow Trettners. In each of these cases, reciprocal services or goods are donated, but no formal reckoning is done, and no one worries much about returning favors or debts outstanding.

Other significant sorts of labor exchange are also carried out in Tret in a more informal manner than in St. Felix. At the hay harvest, when a number of rakers and hay cutters speed completion of the job, villagers will seek help among the members of domestic units with which they already have ties. Labor is donated by those asked if they feel they can afford the time, and they agree to work with no stipulation of any return for their efforts, other than the understanding that they will be provided with food and drink during the time they are working. At a later date, each of the domestic units who assisted will also be seeking help with their own hay harvests, from among those whom they have aided, as well as members of other units with whom they have ties. Although villagers feel an obligation to reciprocate assistance, there is not strict accounting of favors: each household simply draws from the village labor pool and in that way secures the help it needs.

This is not to say that every household can call randomly on every other for assistance. Each has formed a network of relations with other households over the years. Some of these ties are strong, involving almost continuous interaction, whereas with other households interaction is only sporadic, and with still others, rare or nonexistent. The frequency of labor recruitment from other units is but one aspect of the total pattern of social intercourse, and closely parallels the intensity of the total relationship. Although there is a tendency toward clique formation—member units exchanging labor among themselves and excluding outside units—such formations are always incomplete: the labor networks of different domestic units are rarely, if ever, identical. An occasional household, moreover, may be ostracized as the result of a reputation for sharp practice at the expense of other villagers, and excluded from labor exchanges.

Viewed in these economic dimensions, social relations within the two sets of villages have a strong resemblance to Sahlins' categories of generalized and balanced reciprocity:

> Balanced reciprocity is less 'personal' than generalized reciprocity. From our own vantage-point it is 'more economic.' The parties confront each other as distinct economic and social interests. The material side of the transaction is at least as critical as the social: there is more or less precise reckoning, as things given must be covered within some short term. So the pragmatic test of balanced reciprocity becomes an inability to tolerate one-way flows; the relations between people are disrupted by a failure to reciprocate within limited time and equivalence leeways [Sahlins 1965b: 148].

Tret resembles the generalized pattern in the absence of calculation in labor exchanges between both closely related and unrelated house-holds. In St. Felix, such exchanges are of the balanced variety with precise accounting and insistence on equivalence. In Tret, the gen-eralized pattern of labor exchange between particular households tends to be of long duration, often for the lifetime of the participants. This is not so in St. Felix, where such generalized reciprocity is re-stricted to closest kinsmen, and is characteristically short-lived. Even brothers established in separate households within the village care-fully calculate the value of their assistance to each other and demand an equal return. In fact, wary that a sibling might interpret an act of assistance as a brotherly act, which did not require a material return, brothers tend to avoid each other, preferring to call on non-kinsmen for assistance.

General reciprocity in St. Felix is most characteristic of the rela-tionship between a man who is courting and his sweetheart's father and brothers. Although cheerful giving of gifts and loans of equip-ment, as well as donations of labor, characterize the period of court-ship, within a few years after marriage, the father- (or brothers-) in-law will be required to return in kind any assistance received from his daughter's husband or members of his household. Even in Tret there are limits to the practice of generalized reciprocity. Such a relationship is maintained only if it is mutually beneficial. A man would not, for example, donate any of his cattle to a labor exchange partner: labor and objects will be given freely only if their donation does not mean a hardship.

Dealings with domestic groups outside one's village, but still within the three-village area of Tret, St. Felix, and Unser Frau, are based on balanced reciprocity, with the relative values of traditional goods and services understood by all. In hiring a man to work for a period of time or to perform a particular task, only the details of the job are discussed beforehand by the parties involved. Both know how much money the work is worth, and the man who is to do the work assumes he will be paid within a reasonable time after the job is completed—even then the talk rarely turns to money. The pay-ment of money might take place as follows. At the completion of the task the worker is invited into the kitchen for a glass of wine. While drinking the wine, the employer will place the money on the table in front of the employee who will nod, or even seemingly ignore it. When the wine is finished, or perhaps after a second glass, he will stand up to leave, pocket the money without counting it, and be on his way. It is also possible that the employer might not have the cash on hand. If so, the wine is still shared, and if the matter is not brought up by the employer, neither time nor amount of payment will be discussed: the employee will leave without inquiring about his pay. The burden of arranging for payment then lies with the employer, who will take the cash to the employee's house when it is on hand.

Delay in payment may last days, weeks, months, and occasionally more than a year. Only if the employee has an immediate, pressing need for cash will he approach the employer and broach the subject. To demand payment is an unfriendly act and involves considerable embarrassment for both parties. Rather than a direct approach, a man may take an easier route—gossiping about nonpayment with mutual friends. Hearing the gossip, the employer will be moved to initiate payment as rapidly as possible, or face potentially great difficulty recruiting labor in the future.

An alternative way of confronting an employer who has been delinquent in the payment of a just debt is through playing cards, or conversation during a wine drinking bout at a local inn. Drunkenness, provided it is not chronic, can be used to justify unnatural behavior; therefore, the employee may confront his employer in hostile terms, to get across his desire for immediate payment. The employer should visit his house within a few days to make the payment. However, the laborer can also disclaim responsibility for his hostile behavior in having demanded payment, pleading drunkenness. No doubt he will deny a need for the money at the time of payment and even make a show of refusing payment at that time. In the end he will accept, but his protests will set social relations right between the two and mask the hostility underlying the situation.

Economic relations with those outside the three-village area fall generally into the realm of negative reciprocity. Formerly, other neighboring villages, especially those in the Upper Anaunia, were included in the more sociable categories of labor exchange, but the realignment of communication channels produced by the new Gampen road has served to exclude those villages not along the road. With no social or neighborly ties to mitigate the "economic" nature of the transaction, the Upper Anaunian will attempt to make the most favorable deal possible. These villagers share a well-deserved reputation for sharp practice with their brethren from elsewhere in the Val di Non and Val di Sole. Divine intervention is apparently needed by the outsider who would do well in a transaction with villagers from these valleys. The saying, "De Nonensibus et Solondris libera nos Domine" ("preserve us, God, from Nónes and Solandri") goes back to the Middle Ages.

Crosscutting the ideal patterns of labor exchange in the villages of the Upper Anaunia are necessities growing out of the scale of an estate's operation: the ideal patterns work best for the medium and small landholding. On such holdings, the domestic unit contains within itself sufficient labor to handle almost all the work to be done throughout the year. Only a few tasks, particularly the harvesting of grain and hay, the spring plowing, and hauling hay and lumber from field and forest require outside help. Harvesting requires assistance because it must be done in a rush—to leave ripe grain or harvested hay in the field is to court disaster. It is safer to recruit additional labor than to rely on the members of the household alone. On some estates, the seasonal fringe population is adequate to get the job done, but for others an exchange of labor with a neighbor or two is also necessary. Plowing and transport definitely call for neighborly assistance. These chores require a pair of draft animals, and few holdings can afford the luxury of two animals when a single animal will do for all but six or seven days out of the year.

While the fringe population and labor exchange are well suited to the seasonal demands for extra labor on most holdings in the villages, the few holdings of larger size require other arrangements. Managers of the largest estates need assistance more frequently and have less time to devote to labor exchanges themselves. Their greater labor demands can be partly filled by maintaining a larger domestic unit in the form of unmarried siblings or a full-time hired hand, but even this larger force is inadequate to meet the seasonal demands of the harvest. Therefore, individuals are hired on a day-to-day basis: for example, the unmarried men and women living in the village on their natal holdings, managers of dwarf farms, and managers of some small estates.

The small landholder is able to satisfy most or all of his own labor requirements without relying on labor exchanges. Indeed, instead of needing labor at peak seasons, he has labor to offer, and since his holding is not large enough to provide for the needs of his family, he depends on additional sources of income. Some of this income derives from his trade or craft, but he also must rely on periods of employment as a laborer. Thus, there are always hands available when the large landholders need them.

This labor structure tends to even out economic differences within the villages. While the managers of medium and some small estates can meet their labor needs through the mechanism of the labor exchange without sacrificing any of their produce, large landholders must hire wage labor in order to keep all their land in operation. Owners of dwarf estates, in turn, make up for the insufficiency of their holdings by working on the large estates. However, as there are always more laborers available than local demand can utilize, seasonal employment must also be sought outside the villages.

The harvesting of hay on mountain meadows presents special problems because of very low yields per unit area. On comparing the number of hours required to harvest these fields and the value of labor per hour with the commercial value of hay, one can conclude only that the operation is uneconomical. The productivity of mountain meadows varies from field to field as a function of differences in altitude, soil, and moisture. However, a man and a woman working together can expect to harvest 200–300 kilograms of hay in about two days. At 3300 lire per 100 kilograms, this hay would cost 6600–9900 lire if purchased. Working with a scythe, a man would require a full day to cut the hay, the woman a day to rake it. After drying for a few days, the hay must be turned, a half-day's work for each, and in another day or so it must be raked together, put on a wagon, and brought home, another one-half day's work for each. Three hundred kilograms of hay thus would require at least 16 man-hours and 16 woman-hours before it was in the barn. Since the hourly wage for a male agricultural laborer is 500 lire and for a female, 300 lire, the total "market" value of the labor invested is 800 lire per hour times 16 hours, or 12,800 lire. This figure alone is well above the value of the hay on the market, and it does not reflect the cost of hiring a team of horses or oxen for a half day to get the hay home, nor the value of the labor expended in caring for the meadow at other times of the year. It is also worth noting that, by 1969, the value of hay had reached an all-time high, whereas the wage paid for agricultural labor had remained relatively constant for some years.

However, without the hay from these fields, the estate manager would not be able to feed his cattle through the winter. And since he cannot afford to buy hay, he invests whatever labor is necessary to cut the grass on his parcels of mountainside and transport it home. Those with medium or smaller holdings are able to harvest these fields either alone or through labor exchange, since the meadows are scattered at a variety of altitudes and do not all mature at the same time. Only the large landholders face the problem of harvesting more land than can be managed by the domestic work force. Nevertheless, to hire labor at standard wages for this job is out of the question: the problem is solved by holding a hay-mowing bee. The owner of the field to be cut sets the day and word is passed to all the young men and women of the village; on the appointed day, they show up to do the job. In addition to enjoying each other's company, the workers receive food and drink for the day and a small wage. In the evening they return to their employer-host's house, which they take over and convert into a dance hall. Thus, by providing a social occasion, the large landholder is able to obtain the labor needed to harvest his fields, labor he could not afford at the going hourly rate.

The harvest season, beginning in June and lasting into October, is, of all seasons of the year, the most demanding time. The villagers devote virtually every waking moment to their fields. They rise before dawn in order to complete work in the barn and be on their way to the fields at first light. During harvest time, labor exchanges are most important, while there is less time to devote to their ceremonial and social aspects: a minimum amount of time can be spent making the necessary arrangements. Thus, during this season the network of social relations is confined mainly to the villages themselves. Domestic units in Tret draw almost exclusively on each other for assistance, and those in St. Felix, families in St. Felix and Unser Frau. While there is some working back and forth between Tret and the two German-speaking villages, the web of the labor exchanges and the hiring of day labor does not extend outside of the three-village area.

Throughout the rest of the year, villagers from St. Felix and Unser Frau maintain their isolation, entering into social relations with other villagers and outsiders only when necessary for the completion of some particular task. These short-term relationships, typically involving a single interest between the parties involved—the sale of a cow, the sale of hay rakes at a village market, or employment for some period of time in a fruit-processing warehouse—constitute a form of *single-stranded* coalition (Wolf 1966: 81–84). While villagers in Tret also engage in such coalitions, they maintain in addition a variety of *many-stranded coalitions* with neighbors and kinsmen within Tret, and with affines, relatives, and others who live in neighboring villages, in towns and cities within the region, and even with siblings and uncles who have migrated to the Americas. This network of relationships is extended and strengthened by labor and gift exchanges, by social visits back and forth, and by sponsoring each other's "*rites de passage.*" Much effort and time is devoted to keeping such ties alive, especially during periods of relative inactivity in the agricultural cycle.

CHAPTER **VIII**

Inheritance

We have often, in the preceding pages, referred to property inheritance in the context of both village and region. We have seen how inheritance processes developed as an integral part of the Tyrolese political economy, structuring relations between estate managers and territorial elites. At the same time they have patterned social relations within the village itself. As we have seen, the man who manages his resources well is eligible for such tokens of social, political, and ceremonial esteem that the villagers may bestow, and within his own domestic unit he is the dominant figure. Conversely, villagers dependent on resources managed by another cannot reasonably expect to receive recognition and must be content with a subordinate status within the domestic group. To understand both social and property relations in the Upper Anaunia, we must first understand the inheritance process.

Obviously, social and property rights are not static. As individuals proceed through life, the nature of their relationship to each other and to productive resources changes. A child is, with his siblings, subordinate to his father and dependent upon the family estate. As he matures he may stand as heir apparent to his father, and thus become dominant in relationship to his siblings. Still later, he may succeed to management of the estate, take a wife, and start a family. Thereafter, kinsmen who remain dependent on the holding, whether sibling or parent, will be subject to his decisions. As manager of his own estate he will have emerged into what Fortes (1958) has called the "politico-jural" sphere: he will be responsible for the conduct of his domestic unit in the community and eligible for community honors. On the other hand, he may share management with a coheir, remain at home, dependent on a brother who has succeeded to management, or leave the village permanently. In time, the new manager will be succeeded by his own heir or heirs. In this cycle, social and property relations are in a state of continual interdependent development.

However, the developmental cycle of social and property relations within the domestic unit does not proceed in a vacuum. The various domestic units within the community interact, influencing each other's developmental cycle. Moreover, the resulting network of social relations is subject to multiple forces emanating from the ecological setting, the market, and the state.

THE IDEOLOGY OF INHERITANCE

The principal long-term problem of the estate manager centers around the eventual transfer of property to the next generation, a matter that is suggested by his own advancing age, his maturing sons, and his marriageable daughters. One day he will be too old to work, and then he will be dead. Before that happens, he would like to be sure that all his children have been given the best possible start in adult life. He would like to see every daughter well married and every son with land enough to support a family. Then too, he would like to see the holding that he has maintained against the world for a lifetime remain essentially intact to provide a material basis for perpetuation of the family line. However, the meager resources at his disposal are, more often than not, insufficient to fulfill both these goals. He must balance his desires to perpetuate his name against the future of his children.

At one extreme of inheritance possibilities, the perpetuation of the family estate will be given priority over all other considerations— all land and other resources are kept intact and passed on to a single heir. Other offspring are disinherited and left to make their way through life as best they can. At the other extreme, all property, regardless of its extent, will be divided equally among the children. Intermediate patterns, with some degree of estate division and varying degrees of inequality of shares, occur in seemingly endless variations: land may be passed on to a single heir and cash settlements made on the disinherited; land may be divided only among sons, daughters being provided with a dowry at the time of marriage; where a single son inherits the land, other sons may be trained in a trade (see Habakkuk 1955).

However, in deciding what to do with the estate, the peasant is not faced with limitless possibilities: there are guidelines to follow that assign priorities to various factors. The village ideology provides him with a model of how things are "properly" done, whereas a national ideology is expressed in laws and backed by mechanisms of enforcement. National and local views of inheritance may coincide or conflict; although both affect the intergenerational transmission of land rights and other resources, neither one alone, nor both in combination, determines the actual process of inheritance. The use to which ideology is put depends upon the ecologic and economic setting.

Inheritance in the region Trentino–Alto Adige/Tiroler Etschland is regulated by the laws of the Italian state requiring that all of a man's offspring be provided for at his death. Each heir should receive an equal share of each parent's land and other belongings, or else be compensated by a cash settlement equal to the value of his share in the holding. However, some leeway is allowed. While at least two-thirds of a person's property must be divided equally among his offspring, the testator may dispose of up to one-third in any manner he chooses. When the region became a part of Italy in 1918, this law was acceptable in the Trentino, where division of property each generation was already the ideal, but came into conflict with laws in effect when the Tyrol was part of Austria.

While a form of partibility in which a single principal heir received the bulk of the ancestral holding, with a smaller portion divided among remaining siblings, was practiced in parts of the South Tyrol (in Vintschgau and in the wine-producing areas south of Bozen), single-heir inheritance was the ideal elsewhere (Wolf 1970). Impartibility, was encouraged by Tyrolese law from at least as early as 1404. In 1770 and 1785 a special category of impartible estates, "closed holdings" (geschlossene Höfe), was established. Division of such holdings through either inheritance or sale was prohibited, although provisions were made for the free circulation of parcels of plowland and meadow owned by these estates but not part of their "original" composition. Liberalization of inheritance established by Vienna for the entire monarchy in 1868 was countered by a Tyrolese law of 1900 that renewed the acts of earlier years. However, after the absorption of the South Tyrol into Italy and the assumption of power by the Fascists, pressure was brought to bear in favor of partible inheritance.

In 1929, partibility became mandatory, and force was applied to make the regulation effective in the German-speaking regions. Even so, the Tyrolese resisted this effort to abrogate their tradition of impartibility. Under the regime established in northern Italy by the German *Reich* in the last years of World War II, impartibility was reinstituted, and after the war ended, the Italian state did not interfere with its practice. In 1952, the province of Bozen reinstituted the "closed holding" and wrote it into law in 1954. Division of holdings classified as "closed" was again prohibited, and although a number of holdings had lost some land through division in the intervening years, the number of impartible estates in the South Tyrol decreased by only 6% between 1929 and 1954 (Leidlmair 1965b: 570).

This contrast in the national ideology of inheritance is paralleled in the local contrast in ideology between the Tyrolese and Nónes villages: among the Tyrolese villagers, impartible inheritance is the ideal form; the Nónes villagers, on the other hand, prefer the partible inheritance ideology of the Trentino.

In the German villages, impartible inheritance ideally takes the form of primogeniture, in which the eldest son inherits the entire property of his parents, and younger siblings must either leave the property altogether, perhaps receiving cash compensation, or else remain on the estate as dependents in a subservient position. Management of the holding lies in the hands of the principal heir, and all who reside on the holding must abide by his decisions, be they spouse, offspring, sibling, or aged parent. Central to the concept of impartible inheritance is the desire to keep the homestead intact from generation to generation.

Although the bulk of an estate's lands should remain undivided, parcels of land are sometimes attached to the holding and detached later, either through purchase or sale or through inheritance by secondary heirs. This practice of impartibility for the bulk of the land with supplementary parcels of freely circulating land is not only regarded as proper, but it conforms to Tyrolese law. As we noted in Chapter VII, prior to World War I, and again since 1954, the sale of land within each commune has been regulated by the land commission (*Örtliche Höfekommission*).

Nevertheless, the Tyrolese villages in the Anaunia have been under the political jurisdiction of the Trentino, rather than of the South Tyrol, during most of their history: only since 1948 have they been a part of the province of Bozen. The various Tyrolese land laws, enforced only in the German-speaking regions of the Tyrol, consequently did not apply until recently. Although none of the holdings in the German Anaunia was legally classified as *geschlossene Höfe* during the pre-World War II period, the sympathy of the area with the concept is shown by the voluntary declaration in 1954 of fourteen holdings in Unser Frau and one in St. Felix as "closed," as provided for in the legislation of Province Bozen (*Landesgestz*) of that year.

In Tret and the other Nónes villages of the Upper Anaunia the ideal of partible inheritance holds that all the offspring of a landowner should share equally in the inheritance of his homestead. Here the concern is not to maintain a subsistence-producing holding as a constant package through time, but rather to ensure that each child will "have something" with which to begin life. The construction of a living-producing holding comes not from the preservation intact of the holding of one's forefathers passed through an unbroken succession of eldest sons; rather, it is expected that each of the offspring will be able to combine his bit of ground with the bit of ground inherited by his wife and from the combination produce enough land to farm. Thus, within each generation parental estates are broken up and new ones formed from the pieces, the particular pattern depending on who marries whom, and who inherits what.

If followed rigorously, these ideals would lead to extreme outcomes: under impartible inheritance the number of holdings would remain constant through time, as would the composition of these holdings; under partibile inheritance, land would be continually fragmented until each holding became so small as to be economically worthless, and the composition of holdings would vary each generation. In fact, neither outcome has been realized. In German-speaking St. Felix, twenty-three holdings and *Hofrechte* (hereditary rights of use to communal land) are recorded in early documents. Today there are sixty-two holdings. In Romance-speaking Tret the total of land holdings is fifty, and none is so small that it cannot provide a meaningful portion of a family's support; many have shown little change in composition for several generations. In St. Felix, some holdings have been divided, and others have detached parcels either through sale or transfer by inheritance to secondary heirs. Thus, new holdings have been created: traditional homesteads rarely contain all the land held by earlier generations. And in each generation in Tret, some heirs have been disinherited. Out of every group of siblings, one or a few of all the potential heirs have managed to consolidate control of enough land to keep their holding economically viable; others have relinquished their claims or have somehow been deprived of their share of the inheritance. Obviously, then, other factors than the ideology of inheritance must be operating to affect the transmission of property (see the case studies in Appendix 3).

THE REALITIES OF LIFE

In dealing with the inheritance of rights to property, ethnographic reports have usually limited themselves to descriptions of inheritance ideologies. Discussions of the mechanics and sociology of inheritance have been generally lacking. Occasionally the investigator notes that individuals other than those indicated by the ideology do in fact succeed to office or inherit a significant share of the heritable goods, or that there are ways of bypassing the mentally and physically unfit. Systematic attempts to deal with these exceptions, however, remain rare (for one such study, see Gray 1964).

In the Upper Anaunia, both of the prevalent ideologies are honored more in the breach than in the practice. In St. Felix, the eldest son rarely inherits the entire, undivided estate, and in Tret, all siblings do not share equally in the inheritance of the family holding. In St. Felix holdings are divided, or parcels of land detached from time to time; often it is a younger brother, not the eldest son, who inherits all or most of the estate, and at times, all sons may be passed over in favor of a daughter. In Tret it is more usual for one or several heirs to inherit the bulk of the property, while most of the other brothers and sisters either receive but a token settlement or nothing at all.

Certain aspects of life in the Upper Anaunia have made the exact translation of either inheritance ideology into practice virtually impossible. Although these have been discussed in detail in Chapter VI, it might be useful to summarize them here. The subsistence-based economy of the area put a premium on the possession of land. Without some sort of a claim to support from the land it was not possible for individuals to remain in the villages. Furthermore, in each generation more individuals were born than the local economy could support as adults. Careers outside of the area, however, were uncertain and rarely could offer rewards equal to those provided by a village holding.

Table 7

LIFE SITUATIONS OF VILLAGERS BORN 1800–1930[a]

Life situation	St. Felix		Tret	
	No. of men (%)	No. of women (%)	No. of men (%)	No. of women (%)
Remained in the village and married	151 (37.75)	140 (35.4)	123 (34.4)	60 (17.5)
Left the village permanently, after some time in a fringe relationship	139 (34.75)	176 (44.4)	191 (53.3)	248 (72.5)
Remained in the village as a permanent resident or in a fringe relationship	110 (27.50)	80 (20.2)	44 (12.3)	34 (10.0)
Total:	400 (100.0)	396 (100.0)	358 (100.0)	342 (100.0)

[a] Figures include only villagers surviving 20 or more years.

These facts of life are obvious to every villager, who knows that only a percentage of those born in either Tret or St. Felix will be able to remain there as adults (see Table 7), those who inherit a holding or marry a landholder will have the best prospects, and those who must seek careers in the surrounding region face an uncertain future. Consequently, everyone would like to remain on the land with his own row to hoe, and, at least potentially, every member of a sibling set is a competitor to every other for their parents' land. Each generation must be sorted into heirs and disinherited, the inheritance process being as much concerned with denying land to some as in securing it for others. Life strategies collide over the matter of land, and the father with land to bequeath can no more ignore the wishes of his maturing sons than they can disregard the will of their father. The pressures thus generated in the interplay of strategies act upon the way in which property is inherited and can either strengthen or modify the practice of inheritance ideology.

ECOLOGICAL CONSTRAINTS

As a result of the pressures of village life, holdings tend to be confined to a relatively narrow range of sizes: excluding forest land, over 85% of all the holdings within the two villages fall between 0.5

and 10.0 hectares. This narrow range contrasts with the Trentino–Tiroler Etschland as a whole, where holding size varies from less than 0.5 hectare to giant estates of several hundred hectares (Schreiber 1948). When a holding has been subdivided too far, bankruptcy is inevitable. Thus, very small holdings are either combined into larger ones capable of supporting a domestic unit, or are incorporated into existing viable estates. When the holding is large enough to support more than one domestic unit, the landless siblings press for division in order to obtain the material basis upon which they can establish their own households. As a result, almost all of today's holdings are of a size sufficient to support a single domestic unit, but too small for further division into viable fragments.

Often enough, holdings have been divided a bit too far—the division produced two or more holdings, which, while obviously small, were still large enough to tempt the peasant to make a go of it. Three results are possible from such divisions: a man might survive if able to supplement his farm income by engaging in a trade or craft; he might earn enough money to buy more land by working outside the village, or he might enlarge his holding through marriage to a landholding woman; or he might become bankrupt. The lack of alternatives to agriculture has reinforced the practice of keeping holdings from endless fragmentation. Had it been possible, as in the southern reaches of the Anaunia and in the Trentino at large, to supplement income from land through craft industries, for example, to the point that land operation became secondary, then estates could have been divided into extremely small holdings, as has been the case in much of the Trentino (see Chapter IV). But on the mountainside, where such alternatives have been much more limited, villages have attempted to prevent undue fragmentation of estates. Moreover, holdings of marginal size have been continually consolidated into larger holdings.

THE AGE FACTOR IN INHERITANCE

The actual transmission of property rights as well as the kinds of social relationships likely to develop among male siblings, and between offspring and father, is a function of the age difference of heir apparent and father. Other factors being equal, the greater the age difference between father and heir apparent, *the greater the likelihood that he will in fact be the heir; and, conversely, the smaller the age difference, the more likely that the designated heir will not inherit.*

The Tyrolese father is very reluctant to relinquish control of his holding—a situation that usually ends only with his physical disablement or death. It is not unusual to find a man in his seventies or even eighties working vigorously, in full managerial control of the land. Thus, had he married in his thirties, his oldest children are likely to be forty or fifty when he turns the holding over to them. Even if the owner dies at, say sixty-five, his heir in most cases will have spent over twenty adult years working his father's estate. The heir has a long wait, and it is not surprising that friction between the peasant and his heir is common. Despite the clear-cut advantage of patiently waiting it out until one has a holding of one's own, tensions arising from the continuing dominance of the father, and a young man's desires for independence, lead many an heir to forego his inheritance and leave the village. Bitter scenes have accompanied these departures, and the break between father and son may be permanent. Of the thirty-three current holdings in St. Felix where male heirs were available, seventeen of these had foregone their inheritance in this way (see Tables 8 and 9).

As long as they stay on their natal holding, men and women are vulnerable to the whims of their father. He decides what is to be done in the fields on any given day and how to proceed. The purchase of clothing, and even the decision to attend a dance, are ruled on by the parent, to say nothing of permission to spend a day working for another villager, or to seek temporary work as an agricultural laborer in the lowlands. The individual who earns money by his own labor turns over his earnings to his parent.

As long as his father continues to manage the holding, the heir must postpone marriage. Because most holdings are of a size that can support only one domestic unit, the creation of a second household dependent upon the estate is out of the question. The family labor pool is sufficient to work the holding, and more hands simply would mean more mouths to feed.

Thus, the interests of the heir apparent clash with those of his father. Moreover, at twenty-one, a youth is recognized by law to be mature enough to manage his own affairs and take a place as an adult member of the community. Villagers may even confer adult status on younger persons who demonstrate responsibility and competence. To remain at home under the authority of one's father is a sacrifice of independence now for the promise of the material security and social standing that landholding will bring in the future.

Table 8

FATHER–ELDEST SON AGE DIFFERENCE IN RELATION TO
INHERITANCE IN THE NINETEENTH CENTURY

Age difference (years)	St. Felix			Tret		
	Total	Disinherited (%)	Principal heir (%)	Total	Disinherited (%)	Principal heir (%)
<30	21	15 (71.4)	6 (28.6)	27	15 (55.6)	12 (44.4)
31–35	24	16 (66.7)	8 (33.3)	16	8 (50.0)	8 (50.0)
36–40	16	6 (37.5)	10 (62.5)	4	1 (25.0)	3 (75.0)
41–45	4	3 (75.0)	1 (25.0)	4	2 (50.0)	2 (50.0)
>45	6	3 (37.5)	5 (62.5)	3	1 (33.3)	2 (66.7)
Total:	73	43 (58.9)	30 (41.1)	54	27 (50.0)	27 (50.0)

Table 9

FATHER–ELDEST SON AGE DIFFERENCE IN RELATION TO
INHERITANCE IN THE TWENTIETH CENTURY

Age difference (years)	St. Felix			Tret		
	Total	Disinherited (%)	Principal heir (%)	Total	Disinherited (%)	Principal heir (%)
<30	14	11 (78.5)	3 (21.5)	3	3 (100.0)	0 (00.0)
31–35	10	6 (60.0)	4 (40.0)	7	3 (42.9)	4 (57.1)
36–40	11	4 (36.4)	7 (63.6)	6	1 (16.7)	5 (83.3)
41–45	4	1 (25.0)	3 (75.0)	3	1 (33.3)	2 (66.7)
>45	4	1 (25.0)	3 (75.0)	3	0 (00.0)	3 (100.0)
Total:	43	23 (53.5)	20 (46.5)	22	8 (36.4)	14 (63.6)

We have noted that, the eldest son often leaves his father's home and the village. A younger brother—who may be second in line, or even third—is then designated heir. However, younger siblings, with little hope of succeeding to the holding, have often left home to seek their fortune while the eldest son was still at home. They may have made a start for themselves in life, perhaps as craft specialists, and may be reluctant to give up their specialty to return home. Thus, it is often one of the youngest sons who inherits the estate—one who was still young enough to be uncommitted in a life career at the time the eldest made the final break.

To illustrate: on a holding in St. Felix, the third of five sons succeeded to management of the estate. The eldest son, born in 1900, when his father was thirty-five years old, was heir apparent.

He wished to marry a girl from another village holding. The girl's father had no sons, and since the holding was in debt, he saw an opportunity to ensure security for his old age. He agreed to the marriage, provided that his potential son-in-law would agree to pay off his debts and, in time, take over management of the holding. The boy's father opposed the marriage under any conditions: in his early sixties at the time, with young children to support, he was neither willing to turn over estate management to his son nor to support a second household. He was equally unenthusiastic about his son settling on the girl's estate. Friction between father and son grew, and in 1926 the son married the girl and left home.

The father then approached his second son, born in 1903, and offered the estate to him. This son, however, under the assumption that his elder brother would be heir, had apprenticed himself to a blacksmith in Fondo: by the time his brother married, he had been established for two years as the smith in Tret. Despite much pleading by his father, he would not give up his profession. Thus, the father turned to his third son, only twenty-one at the time of the eldest brother's marriage. Thirty-three years old when his father died in 1938, this son inherited the holding.

If a villager marries late in life, when he is near forty, the eldest son will have fewer years to wait between the time he reaches maturity and the death or incapacity of his father. He will be more likely, therefore, to remain at home and assume the rights and responsibilities of the inheritance. In fact, the son who reaches maturity when his father is contemplating retirement, or at his father's death, stands the best chance of inheriting the estate.

Conflicts of interest are sometimes easily resolved: the father may retire from management even though physically able. This is unlikely as long as he has dependent children, but once they are all grown, pressures to turn over the holding mount, especially if his wife is dead. The villagers of both Tyrolese and Nónes origin agree that the goal of a family is to raise children: once this goal has been reached, a man should step aside in favor of his heir. Only if he has a holding can a man marry and start a family. Thus, the father with all adult children who postpones relinquishment of management seems unreasonable to villagers. Unreasonable old men are to be found, but most villagers do in fact turn over their holdings when all their children are grown. The expectation is that they will then take up the role of dependent parent. On occasion, a retired Tyrolese has asserted his independence: sprinkled throughout the histories of various estates in the villages are cases of old widowers who have relinquished their patrimony, only to leave home to marry a widow with a holding and young children and, thus, begin life anew. Two such indomitable souls are currently managing estates in St. Felix and Unser Frau, one having taken up his new life at fifty-three, the other in his sixties.

The above discussion assumes the pattern of impartible inheritance of the German-speaking Anaunia, but it applies equally well to partible inheritance in the Nónes villages. When the father marries late in life, the eldest son tends to assume control of the estate upon his death, employing various strategies to gain control of as much of the holding as possible; younger sons tend to move off the holding and build lives for themselves elsewhere. In contrast, when the father has married young, it is the elder sons who move away and the younger sons who remain on the land. This pattern is reinforced by the tendency of landowners to leave the larger share of their holdings to the offspring who actually remain on the land with them, and to leave a small share to the others. The villagers consider it proper that an individual who remains at home all his life be favored in his father's will.

A case from Tret illustrates both the desertion of the village by elder sons and a unique deviation from the practice of leaving the estate in control of a son who has remained on the holding (see Figure 11). The eldest of the brothers, 2, left home and migrated to the United States in 1909, when he was nineteen and his father fifty-three. The second son, 3, remained in the village until he was twenty-five years old, at which time he, too, went to America. Meanwhile, the eldest daughter, 1, had also left home, marrying a man from Fondo in 1908; another daughter, 6, left the village to marry a low-lander in 1920. That left the third son, 4, and two younger daughters, 5 and 7, at home with their father. However, shortly after 1920 the youngest daughter, 7, departed permanently to work in Meran, thereafter returning home only occasionally on visits and rarely participating in the holding's affairs.

Nonetheless, when the father died in 1939, it was discovered that he had left most of the estate to the absent youngest daughter,

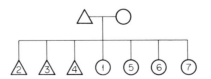

Figure 11.   Genealogy of Tret domestic group.

7, with smaller shares going to the son and daughter, 4 and 5, still at home; token shares went to the three living children established outside of the village. The third son 4, who had expected to inherit a controlling share of the holding, was upset. The father had made no provision to disinherit the daughter should she marry—what if she should marry and return to the village with her husband? Since she owned a controlling share, her brother would be powerless to stop her. She would be able to take over the holding, and he would have to leave the village. In 1948, this sister 7 in fact married a man who was well established in Meran, and would not relinquish ownership although her brother and sister continued to manage the estate. In the same year the brother 4 married and soon after his unmarried sister 5 left to seek employment elsewhere. However, tragedy followed: soon after the birth of a daughter in 1949 his wife died, and the unmarried sister 5 returned home to take her place. The household retained this composition until 1966, when the manager 4 died.

The provisions of this will of 1939 are regarded by the villagers as irregular and "not right." Some felt sympathy for the man, but others were amused at the degree to which the will had placed control over his life in the hands of a woman.

In this particular example, and others as well, a younger son remained on the holding. More frequently, however, the heir is the eldest son. Ironically, the percentage of eldest sons succeeding to management is actually higher than in St. Felix (see Table 10). Since a particular sibling is *not* designated as heir early in life and is not required to remain at home thereafter, elder sons in Tret have more leeway in their life strategies than do their counterparts in St. Felix. They can even leave the village intent upon becoming permanent migrants, and yet return years later to take up life on their natal holding. The father of the current owner of a Tret holding, for example, spent most of his life working in Asia and the Americas. He contributed money to his family, however, and enabled his father to buy land in the village. He eventually returned and became the manager of the estate.

THE INHERITANCE PROCESS IN TRET

The preceding discussion has established the general outline of the inheritance process in the Upper Anaunia; we shall now focus on its separate mechanics in Tret and St. Felix. Throughout the re-

Table 10
SOURCES OF MANAGERIAL RIGHTS TO CURRENT ESTATES

St. Felix		Tret	
I. St. Felix holdings obtained through inheritance.	51	I. Tret holdings obtained through inheritance.	31
A. Inheritance of estates that were not divided.	41	A. Inheritance of estates that were not divided.	25
1. Only son succeeded father as manager.	8	1. Only son succeeded father as manager.	5
2. Eldest of several sons succeeded father as manager.	2	2. Eldest of several sons succeeded father as manager.	3
3. Only son or eldest son succeeded to management of estate owned by mother.	0	3. Only son or eldest son succeeded to management of estate owned by his mother.	0
4. Son other than eldest son succeeded father as manager.	8	4. Son other than eldest succeeded father as manager.	5
5. Son other than eldest succeeded to management of estate owned by mother.	3	5. Son other than eldest succeeded to management of holding owned by mother.	0
6. Manager inherited managerial rights from other than a parent (from a guardian, five cases; from mother's brother, one case).	6	6. Manager obtained managerial rights from other than parent (male heir—from a brother).	1
7. A daughter inherited the estate because there were no available male siblings.	6	7. A daughter inherited the estate because there were no available male siblings.	0
8. A daughter inherited the estate from a parent with one or more male siblings passed over.	2	8. A daughter inherited the estate from a parent, with one or more male siblings passed over.	1
9. A female owner inherited the estate from other than a parent.	1	9. A female owner inherited the holding from other than a parent (one from husband's siblings; one from deceased husband's father).	2
10. Management by a sibling set that includes the eldest brother.	1	10. Management by a sibling set that includes the eldest brother.	4
11. Management by a sibling set that does not include the eldest brother.	4	11. Management by a sibling set that does not include the eldest brother.	4
B. Inheritance of holdings that were divided.	10	B. Inheritance of estates that were divided.	6
1. Managed by eldest son; inherited from father.	4	1. Managed by eldest son; inherited from father.	2
2. Managed by son other than eldest; inherited from father.	1	2. Managed by son other than	

(continued)

Table 10 (continued)

St. Felix		Tret	
3. Ownership inherited by daughter; inheritance from father.	2	the eldest; inherited from father.	3
4. Inherited from other than a parent.	3	3. Manager obtained estate from other than father.	1
		II. Tret holding obtained by other than inheritance.	19
II. St. Felix holdings obtained by other than inheritance.	11	A. Traditional estate obtained by purchase.	8
A. Traditional estate obtained by purchase.	2	1. Holding obtained intact.	5
1. Holding obtained intact.	2	2. Holding divided into two estates, each obtained by purchase.	3
2. Holding divided, each obtained by purchase.	0		
B. Composite estates.	7	B. Composite estates.	9
C. Manager rents the estate.	1	C. Manager rents the estate.	1
D. Other.	1	D. Other.	1
Total number of estates in St. Felix:	62	Total number of estates in Tret:	50

mainder of the chapter we will employ the present tense, even though the opening of economic alternatives since 1950 has somewhat modified the stark picture of an economy characterized by a Malthusian relation between population and land. We feel justified in placing the discussion in the "ethnographic present," since much of the economy remains oriented toward subsistence; hence, traditional rules apply. Even in the case of estates now increasingly devoted to production for market, however, the past mechanics of the inheritance process are still at work.

In Tret, as we have seen, the life goal of the estate manager to provide each of his offspring with a good start in adult life ideally means that each son has his own land and each daughter a husband with land, but the modest scale of his operation makes this goal unattainable. He knows that unless he has few children, he will not be able to provide for all. Likewise, his children realize that only some of them will be able to remain on the land; others will be forced to seek careers elsewhere.

When a man dies without a will, or when a will is drawn up while all his children are still minors, each child will be awarded an equal share of his property. However, since the primary goal of the father is to see all his children established in life, a will drawn up after some of his sons and daughters have reached adulthood may be modified to reflect differing degrees of independence from the

natal homestead. In this, as we have seen, Italian law is permissive—
it does not insist upon an actual division of land, but permits com-
pensatory cash payments; it also permits up to one-third of the
property to be distributed as the owner wishes. Thus, wills may be
used to favor those who are still in some degree dependent on the
holding at the time the testament is drawn up. The peasant may
leave only a token share of ownership to married daughters and to
other children who are established in life; or, he may leave them a
full share but specify that those who remain in the village have the
option of making them a cash settlement. The larger share of ownership then is divided among the children who have stayed in and
around the village. However, it is rare for anyone to be left out of
the inheritance entirely. Not only does the will transmit the means to
a livelihood, but it also expresses the parent's interest in the well-
being of the heir. To be totally excluded from inheritance is to be
disinherited in the most severe sense of the term.

However, if any offspring are to receive land enough to provide
a living, most heirs *must* be deprived of land. To follow the national
inheritance ideology is a practical impossibility. Ways must be found
to concentrate management of the holding in the hands of the number of the heirs it can support. In extreme cases, parents have "sold"
all or most of their estate to the intended heir. Such a transaction
involves little more than a transfer of title since no money actually
changes hands, thus eliminating the need for a will. But most men
do not wish to cede control of the land while they are alive, and most
estates are passed on through a will or the provisions of intestacy.
Because Italian law makes no distinctions among different kinds of
property, the testator can specify which heirs are to receive title to
all (or most) of the land, with the provision that they pay the remaining heirs an amount of cash equal to the value of the land and other
resources they are due. Thus, an estate manager in Tret, who died in
1939, specified in his will that his six children, one male and five
females, were to each receive one-sixth of his estate. In fact, however, the son was to inherit the house, and all the land, livestock, and
equipment—but he was obligated to compensate each sister in cash
to the value of one-sixth of the estate at such time as he was able.
Should he die before this obligation was met, his own heirs would
inherit both the property and the debt. Thus, the letter of the law
is satisfied, but actual control over the property is limited to a specific
heir or group of heirs.

In the above example, the actual heir to management of the
estate was determined by the will, but this is not usually the case.

When an owner dies intestate, or simply specifies that a number of his offspring are to share in ownership of the estate, the problem of succession to management is unresolved. The recording of the names of the several new owners in the commune archives satisfies the state, but merely removes the problem of settlements from the politico-jural domain into the domestic. Two or more heirs may hold equal rights to an estate on record, but, as the result of developments within the domestic unit, one heir may be enjoying full control over the estate, while others are totally or partially excluded from participation in its affairs. How the heirs manage their shares is left up to them—it is never specified in their parents' wills. At least initially they will continue to operate it cooperatively, as they did before their father retired or died, but as time passes, pressures to actually divide the land will mount.

The most immediate demands for a change in the status quo are likely to come from siblings who have married into other village holdings. As they no longer receive any support from their natal holding, they will be anxious to translate their share into something tangible—by selling to the owner or by exchanging rights in the whole for complete ownership of a part, usually a parcel or two of land. The preference of these siblings is clear: they would like the land.

In this subsistence economy, money to purchase an ownership share is not likely to be on hand, and a piece of land now is worth far more than a promise of money in the future. The claimant is especially likely to be successful if he is a brother who has married a woman with land. Sisters, however, may have to make do with a promise. Fathers, whose goal it is to see all their offspring established in life, regard a girl's marriage to a landholder as a fulfillment of this goal. For her to receive a part of his holding as well would be to favor her over the other children. Therefore, wills specifying that daughters who are married at the time the will is drawn up are to receive cash settlements also frequently state that any single daughters who inherit a share in the holding are to be paid off if they subsequently marry. Even when this provision is not actually stated, those remaining on the land are unlikely to allow a sister who marries to have land unless her need is great. If her husband's holding is so small that her life with him is likely to be a real hardship, they may allow her to have a few additional parcels of land, especially if their own holding is relatively large. The full range of possible complications discussed here is rarely met in any single instance, but most property transmissions produce some problems and, for one

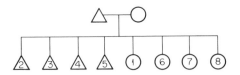

Figure 12. Genealogy of Tret domestic group.

reason or another, several parcels of land are likely to be detached from the parental holding.

While the field study was in progress, one of the largest holdings in Tret was in the middle of such a conflict. At the time of the father's death in 1961, two daughters, 6 and 7 (see Figure 12), had already left home, both migrants to America, and the second son 3 had left the village to work in the Trentino. The eldest daughter was married to an estate manager in Tret, and the eldest son had laid claim to a second building complex owned by the holding and married a girl with an estate of her own. Two brothers and a sister, 3, 4, and 8, remained on the holding.

The father had specified that each of his offspring receive an equal share in the holding, and almost immediately the two siblings who were married in the village, 1 and 2, began to press for division of the land: both wished to translate their share in the undivided holding for one-eighth of the land. The husband of the eldest daughter (1), was an active entrepreneur, one of the first in the village to turn his efforts toward commercial milk production. For him, the fields his wife would receive meant meadow enough for another cow or two in the stall, and that meant more milk to sell.

The married brother 2 needed the land for a different, but related, reason: his wife's estate was too small to support his family, and in order to make a go of it he had taken a job with a neighboring commune, working as a laborer on a project to build a community water system. He commuted daily to this job on his motorbike, leaving before sunrise in the morning and returning late at night. He hoped his share of the estate, when added to his wife's land, would enable him to give up the outside job and support his family from the land alone.

The siblings on the holding recognized the legitimacy of these claims, but wished to postpone the actual division as long as possible—any loss of land from the estate would mean a reduction in their own income. They successfully resisted division through the 1966 agricultural season, but capitulated in the winter of that year.

They continued to manage five-eighths of the estate, their own shares plus those of siblings 5 and 6. One-eighth of the holding was detached by the eldest daughter (1), and the married brother took over control of the remaining one-fourth, his own share, plus the share of a sister 7, whom he was obligated to compensate in cash.

In the final settlement, the two brothers who remained at home succeeded in protecting themselves against further division. One of the sisters living in America (6) returned for a visit during the summer and asked that her share be given to the youngest sister (8) when the division was made. However, the two brothers refused to do this: instead they registered her as a part-owner of the five-eights share they manage. Had they honored her request, and had the youngest daughter someday married, she would be able to claim one-fourth of the estate, instead of the one-eighth she inherited.

Once an estate has been transmitted through inheritance and some parcels perhaps detached, the remaining land may be managed in several different ways. The holding may be kept intact and managed jointly by the remaining siblings for some time, as in the example above, or even for a lifetime. About 10% of the village holdings have been run by such domestic units at all times in the past for which records are available. Since alternatives to remaining on the holding are limited, the number of siblings who stay at home tends to approach the maximum the land will support. The marriage of any man on the holding is therefore a threat to the security of the others. And since all are very directly affected by such a marriage, the courting process involves not only the mutual acceptance of bride and groom, but approval of the bride by the entire sibling subset. Long courtships result, during which the bride-to-be is gradually integrated into the social life and work routine of the household. The plans may come to naught, however, if frictions develop, as they often do—especially between the girl and her fiancé's sister. If such tensions can be avoided until after the marriage, by a diplomatic girl, or resolved by a man willing to put his sisters "in their place," the marriage will finally take place.

Even after the marriage, a holding will be kept intact, the unmarried siblings cooperating with the married couple in its management. However, the addition of a spouse, together with the family raised by the newly married couple, will in time crowd at least some of the siblings into the fringe population; if one or two remain at home, they will tend to treat the married sibling as the principal owner of the holding because his children will succeed to its management.

Since most holdings are only large enough to support a single domestic unit, further division is unlikely, and, thus, only one heir dependent on a holding may marry. However, on the few estates of larger size, the possibility always exists that another brother may seek a division in order to obtain land to support a family of his own. Since both brothers have equal claim to the estate, division is likely. In the division, the ownership shares of the unmarried siblings must be dealt with, and all parties in some way involved with the estate, consulted.

Thus, while the inheritance process tends to produce a single individual in managerial control of each holding, the possibility always exists that this control will be shared with one or two unmarried siblings who remain on the estate. Almost certainly there will be unmarried siblings in a fringe relationship, and still others who own shares of the holding, but who have migrated, joined the church, or otherwise become completely independent of the natal homestead. This situation produces an awkward separation between managerial control and ownership of land. Those who remain at home will have full control over the disposition of the fruits of the land, but they cannot mortgage or sell it without the cooperation of all of the part owners, wherever they may be. On the other hand, ownership of a share of a holding does not automatically confer rights to a living from the land: the right to support is earned only by active participation in the operation of the estate. Siblings who absent themselves completely from the holding lose all rights to an income from the land; siblings in a fringe relationship, however, retain the right to periodic support. Armed with the ownership of a fraction of the holding, this right is secured with continuing donations of labor and cash.

The problems caused by the separation of management and ownership sooner or later bring the manager to attempt to consolidate ownership. If not himself, then his children should have free title to the land—with no possible interference from distant cousins. Unmarried siblings of the manager present the smallest problem. Even if they will not give or sell their share to him, they can be counted on to will them either to the manager or his children. An heir who has left the village and married presents a greater problem. If the manager does not obtain his share of ownership prior to the death of the absent part-owner, they will be inherited by the children of the absentee owner. In a generation or two, it is possible for a fraction of an estate to have literally dozens of owners scattered over a continent or two, some of them perhaps unaware of an ancestor who came from a place called Tret. Such cases have for-tunately not been frequent, but they have occurred.

If an estate manager is lucky, the siblings who have left home and prospered will donate their ownership shares to him, as will siblings who have become priests, nuns, or monks and renounced worldly goods. Most are not so lucky. Usually the money to pay the cash settlement must be earned from the holding itself, supple-mented, in all likelihood, by laboring outside the village. However, on subsistence holdings cash income is meager, and generations may pass before consolidation of ownership is completed. The ac-cumulation of the necessary money can become the major economic burden in the landholder's life.

Thus, despite an ideology of partibility, the ecological situation and the marginal economic position of the village prevent the per-petual division of holdings in Tret. Furthermore, each generation must reduce the number of dependents on the estate from an entire sibling set to a single individual. Which brother ends up in control of a specific holding cannot be predicted beforehand, but, as we have seen, his selection is not a random process.

THE INHERITANCE PROCESS IN ST. FELIX

Since the ideology in St. Felix is to maintain each holding intact through the years, the problem is not to explain what has prevented fragmentation, but rather to account for the division of property that had occurred. We have already seen that holdings have, in fact, been divided, and despite the differences in ideology between the two villages, both the mean holding size (4–5 hectares in each village) and the range of holding sizes are very similar in Tret and St. Felix. Furthermore, we have noted that in the absence of reasonable alter-natives, younger brothers are likely to put pressure on their father and his heir to divide the holding, so that they too can obtain the means to support a family.

While the father designates a single son as his successor and heir, he does not completely eliminate all his other children from consideration in his will. He would like to provide all with a start in adult life, insofar as he can without impairing the unity of the homestead. As in Tret, he considers the position in life of each of his offspring in drawing up his will. Married daughters, as well as children who have migrated, joined the church, or otherwise established themselves outside the village, can be excluded completely, whereas those still at home or in a fringe relationship to the holding continue to be a source of parental concern. This concern can be expressed by awarding them rights to a living from the land, either written into the will or simply made clear to the principal heir as a parental desire.

When such concern is expressed in a will, the secondary heirs are typically awarded ownership of a room in the house and usufruct of the land. This apparently secures them a living for as long as they remain at home, but they are clearly in a position of subordination vis-à-vis the principal heir, with little voice in the operation of the holding. When their rights are based on no more than the desire of a parent, their position is even more tenuous—in fact, entirely dependent on the good will of the inheriting brother. In either case, the manager of the holding can unilaterally decide their fate: he has a free hand where his siblings' rights are not protected by a will. Even when provisions have been made in a will, their rights can be terminated by a cash payment whenever the heir wishes. It is also understood by all parties that women lose their usufruct at marriage—whether this is specified in the will or not— as does any individual, regardless of sex, who subsequently leaves the village and ceases to participate in the estate's operation.

Since 1919, when the Trentino–Tiroler Etschland became part of Italy, and especially after the Fascist assumption of power, the official policy of the state required that Tyrolese holdings be divided. This edict was resisted in St. Felix, as it was throughout the Tyrol. Various means were used to evade the law, such as selling all or most of the estate to one son while the father still lived, in a transaction in which no money changed hands. More often, the Tyrolese evaded the law simply by outliving the Fascist state. Yet even when the father died, and all his children were duly recorded as owning a fractional share of the estate, the land was rarely divided. The ecological position of the villages was little different in the 1930s and 1940s than it had been earlier, and there was no way for a person to support himself on a fraction of an estate. A single individual would succeed to management of the estate, as before, and would strive to consolidate the estate much in the same manner as managers in Tret—by paying off some heirs, receiving shares from others as gifts, and relying on unmarried siblings to will their shares to himself or his children. A recalcitrant individual might have appealed to the Italian courts and won a share of the land for himself (villagers have never been shy about going to court); but, thereafter, he would have to live with the stigma of being a collaborator. No record exists to show that anyone in St. Felix has ever taken this course. There was some division of property under Fascist rule, but most had taken place prior to that period. In 1879, the number of holdings had grown to fifty-six, only a few less than the sixty-two domestic units that exist today.

The division of estates has occurred on holdings large enough to suffer the loss of a number of land parcels without seriously impairing the potential to support a domestic group. These parcels consist, for the most part, of forest or mountain meadow, but they could also include plowland and village meadows not part of the traditional holding—lands purchased or brought into the holding by the marriage of a former owner. If the main holding is large enough to support the principal heir and his dependents, some of these lands could be sacrificed in the interest of a sibling, either male or female, with prospects of marriage to a village landowner. Such a move is particularly likely when the prospective spouse's holding is small, and the additional land could make the difference between a marginal and an adequate existence.

On estates large enough to support more than one domestic unit, younger brothers are likely to press for division, and fathers are unlikely to resist such demands. A large holding needs more labor than can be supplied by the nuclear family, especially when the children are young. Typically, labor is provided by the siblings who remain at home, or by an outsider who receives room and board in return for his labor. However, in St. Felix, the greater output of a larger holding is almost impossible to translate into a higher standard of living: one can eat only so many potatoes and so much sauerkraut, and marketing surplus produce over mountain trails is unprofitable. At best a few more head of cattle might be raised and sold, but such holdings cannot be commercialized. A large holding may provide subsistence, but not profits. Thus, division of the estate is unlikely to have much effect on the material well-being of the principal heir. Instead of keeping a brother and sister attached to the household of the principal heir, why not create two separate households, and if two households can be supported, why not let the second be composed of a younger brother and his wife? Such pressures have been difficult to resist through the years. Even where documentary evidence is lacking, it is possible to reconstruct the outlines of the ancestral holdings in St. Felix, and to show how the ancestral lines of current holders of neighboring estates converge upon a single ancestor within the last two hundred years. As of 1965, only one holding had escaped major division of its original lands; yet even this estate was divided during the period of field-work.

Where large holdings can be divided into smaller estates, each one with all the needed resources, such a division is impossible in the case of medium-sized or small estates. The owners of smaller holdings have therefore tenaciously resisted such divisions. However, the lands crucial to a holding are its plowland and village meadows, where all of the crops and most of the fodder for the animals are produced. Outlying land, the forests and mountain meadows, play a supplemental role, although they comprise the greater part of most holdings. In a number of cases, secondary heirs have exploited this fact and have convinced their father to provide them with several contiguous parcels of mountain meadow or forest, which they then have attempted to convert into plowland and meadow. Small in size and located on land marginal to the use to which it is put, such holdings have produced a poor living and not all have survived.

INHERITANCE AND WOMEN

Thus far, discussion has centered on the inheritance of property by men. Women frequently inherit or purchase land, but they become principal heirs only under circumstances that villagers regard as extraordinary. The expectation for most women who remain in the villages is to be provided for either through marriage, or, if unmarried, through subsistence rights to their natal homesteads.

In both St. Felix and Tret, parents hope that at least some of their daughters will marry. Since a man without land cannot hope to support a family, a woman should marry a man with a holding of his own. Marriage for a woman is regarded by all as the equivalent of succession to management for a man. It establishes her right to support from her husband's estate. At marriage she is therefore required to give up her claim to support from her father's land. When marriage takes her outside the village, she loses all claim to support from village lands. At most, she receives a cash settlement for whatever share of her father's land she held. However, if she marries another villager, she may be able to exchange her fractional share of ownership in the natal holding for the complete ownership of one or more parcels of land. She and her husband will then work these parcels together with his land. It is unlikely that she will receive these parcels at the time of her marriage, as she must wait until her father dies or retires. Even then, the actual division of the estate will depend on the decision of the principal heir (St. Felix) or heirs (Tret). Any land that a woman receives is owned by her, not her husband. It therefore can be sold only with her approval, and it is transmitted to her children by her will, separately from the land owned by her husband. Whether a woman marrying within the village actually receives land is largely a function of the relative sizes of the holdings of her husband and her father: it is most likely when she and her husband have a real need for more land and her natal estate can withstand the loss.

Marriage thus furnishes a woman with a livelihood. A dowry is one of the prerequisites for marriage, but it is not considered to be a substitute for inheritance. The couple must have the furnishings for their apartment in order to marry, and the bride traditionally provides certain of these, but, they occasionally are provided by the groom. If he can afford them and his bride's family cannot, he will not stand on ceremony—better to let the wedding take place—he had waited long enough as it is. Even when a bride brings a dowry, all or part of it very likely has been supplied by the woman herself, rather than her family. Virtually every woman spends a part of every year working away from the village once she has finished school, and any money she saves will be invested in her dowry.

Not all of the daughters of the Upper Anaunia marry, but this is not regarded as tragic as long as they are provided for in some other way. Some become nuns; others succeed in finding careers as barmaids or housekeepers in the city. In St. Felix, women who have become independent will be excluded entirely from the inheritance, but women who remain at home, or are in a fringe relationship, will be provided with the right to a living, with usufruct, as we have seen above. In Tret, it is more likely that all unmarried daughters will be included in the inheritance, each receiving an appropriate share of ownership in the parental lands. As long as they remain single and continue to participate in the operation of the holding, their claim to support from the land remains intact. In fact, since their shares in the holding may well be as large as those of their brothers, women are able to share in managerial decision-making in Tret. In both villages, sisters who remain on the holding serve as domestic managers in lieu of a wife as long as the estate manager stays single. As these women are invariably reluctant to give up this role, there will be friction when he decides to marry.

Single women thus retain some degree of material security, a security they are required to exchange at marriage for support from a husband's holding. A married woman's rights are in some ways less secure than those of an unmarried woman who remains on the natal homestead. Should her husband die, her children will inherit the holding: she obtains the right to a living from the land, but no legal control over its sale or operation. Should her husband die before any children are born, the land will be inherited by his siblings, and her claim even to usufruct will be tenuous at best. A number of informants expressed the opinion that under such conditions a widow would have no legitimate claim even to a living from the land.

As we have seen, there are cases in which a woman becomes principal heir. This always happens when a man has only daughters: in St. Felix, thirteen holdings are currently held by women who inherited the property from their parents, and in Tret the number is three (see Table 10). While a couple will rarely pass over a son in favor of a daughter, this has happened in situations where the holding is deeply in debt. Either the sons may decline to accept a holding threatened by debt, or the father may decide in favor of a daughter who is being courted by a young man willing to accept the debt or able to pay it off immediately; migrants returning from the Americas with bulging billfolds have, on several occasions, been able to obtain holdings under such circumstances. When the debt is paid off, the woman's father may transfer estate ownership to his new son-in-law. In other cases, the daughter may inherit the land, her husband thereby obtaining a propertied wife; alternatively, a compromise may be struck in which man and wife share in the ownership.

SECONDARY HEIRS

In every generation some siblings become owners or managers of land; the other siblings must yield to them, *weichen*, as the Tyrolese say. In St. Felix, the inheritance process typically establishes a single owner-operator for each holding; all other siblings lose their legal right to support from the estate. Subsequent claims to any such support are based upon their particular kinship ties to the landowner, rather than legal claims enforceable by law. Whether a claim will be honored depends upon the decision of the land-owner. His first consideration is the material well-being of his own nuclear family; the support of his siblings runs a poor second. When the owner-operator requires additional labor, he may encourage a few of his siblings to remain on the estate and remunerate their work through payments in kind. Since holdings in St. Felix rarely admit the support of a unit larger than the nuclear family, fully dependent adult siblings are few in number.

More numerous are the fringe members of the community—men and women who seek a living outside the villages, but who frequently are forced by circumstance to seek at least partial support from their natal homestead. Since even on small holdings there is a recurrent need for additional labor, the heir is frequently willing to countenance a continued relationship with these disadvantaged siblings. He will provide limited support in exchange for occasional donations of cash or work. Yet, even in such cases, the well-being of his own family always takes precedence over such secondary claims.

In Tret, patterns of partible inheritance do not divide the sibling set into a principal heir and dependent siblings. Siblings who remain in the village at the time of the father's death usually share equally in the legal ownership of the holding and take part in operational decision-making. Others who have gone away to work in the lowlands also receive a share of ownership that can be translated into a claim to support from the siblings operating the estate: their right to this support does not depend upon their personal relations with the manager, but is a *legal* right stemming from their part ownership. In practice, however, they can keep this legal right operational only by well-timed donations of labor and cash. If individuals fail to return periodically to the village to extend such aid, their legal rights lapse. The only exceptions to this rule are villagers who return after long absences abroad, equipped with a sizable roll of money. Some already may have made gifts of cash from abroad; others revalidate their rights to support by making a portion of their money available to the estate manager upon their return.

Individuals who work the land full time and those who migrate but remain in a fringe relationship to the holding share a common interest in keeping the holding undivided. The heirs who remain prefer to keep the estate intact in order to have the largest possible holding; the migrants with ties to their natal estates also prefer to see it remain undivided, since division would threaten its capacity to furnish occasional room and board. While some villagers may prefer to buy out their absent siblings, others are content to allow them to retain partial ownership: it ensures a continued interest in the success of the operation and, hence, encourages donations of labor and cash. It is also expected that these individuals will leave their share to the children of the estate household, since they themselves are rarely able to marry and to have families of their own.

On the other hand, the interests of villagers who continue to live on the land and those of siblings in a fringe relationship are often opposed by married brothers and sisters who remain in the village on their spouse's homestead. These siblings prefer an actual division of property in order to convert a fractional share of ownership into outright ownership of several plots that can be worked together with their spouse's land. Daughters may be bound, under the terms of a will, to accept cash in exchange for ownership rights upon marriage. However, sons fortunate enough to marry a village woman with land invariably demand ownership of land parcels in return for giving up any claim to the entire holding. Despite considerable pressure from siblings residing in the village, those who remain on the holding are often able to resist division for remarkably long periods of time.

Conflict of interest among siblings is a major source of familial friction, particularly when continued for more than one generation. The heirs of the man who actually worked the land feel they have a right to retain ownership of the entire holding and that they have only to make a cash settlement on all part owners. Their cousins might well argue otherwise and insist upon a division of property so that they can acquire certain parcels of land.

SUMMARY

It is clear that the ideology and the reality of inheritance in the Upper Anaunia contradict one another. In spite of differing ideologies, the size of holdings and the composition of the domestic unit are remarkably similar in Tret and St. Felix. In Tret, ideology supports the division of larger estates into family-sized holdings. When a holding cannot be divided and remain viable, however, ideology and reality are in obvious contradiction, an opposition that can be resolved only by a process that tends to eliminate all but one of the potential heirs. In St. Felix, a single heir is the ideal; the eldest son should be designated heir, thus maintaining optimum-sized holdings under the management of a single individual. But here, too, there is a contradiction in that ecological pressures tend to force the division of larger estates, and on all holdings—small or large—these pressures often select an heir other than the one ideally suitable.

Nevertheless, the ideology plays a part in resolving the very contradictions it helps to create: it does so by assisting in the definition of personal priorities. The two life goals of the land-holder—to maintain the family estate intact and at the same time to provide for all of his children—are clearly opposed to each other. It is rarely possible for a man to accomplish both; hence, priority must be assigned to one goal over the other. In Tret, the ideal solution is to sacrifice the continuity of the estate to make the maximum possible provision for each offspring; in St. Felix, it is to sacrifice the well-being of secondary heirs to keep an estate intact in the hands of the single heir. In the Upper Anaunia, neither ideology can be fully realized, and every attempt to apply ideal forms within eco-logical constraints results in new contradictions. Thus, while the inheritance ideology provides a cognitive framework within which the de facto process must operate, both the mechanics of the process and its results are in the last instance determined by the forces of environment and market, and in spite of ideologies.

While ecological processes relate directly to inheritance prac-tices, the relation between inheritance ideology and inheritance practice is more circuitous. The analysis of this relation will require several additional steps.

Our discussion of inheritance has focused on the transmission from owner to heir, from generation to generation. The process of property transmission, however, is transgenerational in a larger sense. In each generation, the process produces two important out-comes: it secures the continuity upon the land of some offspring, and relegates others to marginal positions. Generation after genera-tion, some members of a kinship unit are sacrificed in order to ensure the perpetuation of corporate entities made up of successive domestic groups. In such a corporate entity:

> The rights and obligations which attached to the deceased head of the house would attach, without breach of continuity to his successor; for, in point of fact, they would be the rights and obligations of the family, and the family had the distinctive characteristic of a corporation—that it never died [Maine 1963: 179].

In accomplishing this end, however, the inheritance process works simultaneously at two levels: at the level of the domestic group and at the level of the village as a whole, in the public domain. On the level of the domestic group, it sorts a sibling set into heirs, the disadvantaged, and the disinherited. On the level of the village, the heirs emerge as the anchormen in social exchange. As owners and managers, they are eligible to marry, to engage in labor exchanges, and to represent the domestic group in public affairs. All other villagers are subject to their decision-making and relegated to subordinate positions. Purely domestic in appearance, the inheritance process thus has vital public dimension: it continuously creates and recreates the conditions for the recruitment of the village core population, while creating the conditions for the development of a village fringe population and for the expulsion of supernumaries. Above all, it creates a local "elite"—however humble in comparison with the elites of the world beyond the valley—an elite in control of the means of production. And it provides the elite with a labor force of disadvantaged siblings who work in exchange for little more than subsistence wages. The inheritance process is thus the key force in creating both the village labor market and the village order of social precedence.

At the same time that it creates a labor pool, however, it also inhibits any awareness of a common fate among the village landless and marginals. The disinherited and disadvantaged villagers remain attached to the ancestral holding as a source of possible support in adverse times. When such times befall them, each of them has to enter into separate and individual relations with their more fortunate sibling. The landless do not come to form a class *für sich*, as Marx would have put it, a class conscious of its common situation and hence aware of common interests. The relation between employer and employee remains familial, limited to transactions between close kin within the domestic sphere. The inheritance process thus not only produces public outcomes, in dividing villages into village core and fringe, but also changes these public outcomes back into domestic transactions.

We are now better prepared to discuss the role of inheritance ideology. Inheritance practices organize domestic units in ways that produce public outcomes; but the public outcomes affect the distribution of power and influence in the villages, and hence bear directly upon social and political relations within the villages and without. Therefore, inheritance ideology cannot be understood with reference only to local ecological processes and domestic groups: it bears upon village organization as a whole. Thus, we shall again look at the relation between ideology and practice, after we have discussed the patterns of social organization in the villages and their political commitments (see Chapters X and XI).

CHAPTER IX

The New Economic Order

The production strategies followed today in Tret, St. Felix, and the other villages of the Upper Anaunia are the result of past emphasis on subsistence production and the villagers' perception of current and future opportunities for the sale of produce. Before 1950 the Upper Anaunia was involved only marginally in the commercialization of Tyrolese agriculture. Even during the nineteenth century and the first half of the twentieth when production in lowland and middle-altitude communities became specialized to meet the demands of a growing market, the Upper Anaunia, like other upland valleys, continued to raise the full range of traditional crops and animals. Cattle were sometimes driven to market in Malé or Bozen, lumber was carted to sawmills with access to the rail head in Fondo, and a few objects manufactured at home were back-packed and sold at local markets in the lowlands, or peddled from house to house. But the cash income from market sales was in no way central to the economic planning of the Anaunian peasant. He hoped only to earn money to pay local specialists, to satisfy the demands of state and church, and to meet legal or medical emergencies. For

"Cattle were sometimes driven to market in Malé or Bozen." Cattle market, Bozen.

most peasants to sell their goods even occasionally was, at best, an inconvenient necessity.

It is true that the produce of the Upper Anaunia was in demand: grain, potatoes, meat, and milk products were all required by townsmen and by specialized agriculturalists in the lowlands. However, the limited scale of his operation, the relatively poor quality of his produce, and the difficulty of transporting goods to market combined to discourage the peasant from making the transition to market production. Only during the past twenty years have there been sufficient pressure and opportunity to induce local estate managers to switch from subsistence to market strategies, to rationalize production, and measure the income of their estates against the income to be derived from other economic opportunities.

CHANGING CIRCUMSTANCES

In discussing the changes that have taken place in their way of life, villagers usually mention the asphalt road "Mussolini built" through the area in 1936. The road connected the villages of St. Felix, Unser Frau, and Tret with Fondo and with other parts of the Anaunia to the south, and with the population and market centers of the Etsch valley over the Gampen Pass to the north. Daily bus service was initiated over the route, and trucking services from lowland cities became an alternative to horse-drawn wagons and peddling on foot. The Upper Anaunia thus was brought closer to the markets around it; transportation and physical isolation no longer presented the obstacles they had earlier. Access to markets, however, was in itself not enough to bring about an immediate increase in market production, for the simple reason that the area had nothing new to offer. The turbulence of European affairs during the fifteen years following 1936 touched the life of every villager, but it initiated only a few changes in economic organization. The main effect of this period was to draw manpower from the villages. In St. Felix, the *Optionzahlung* of 1939 saw the migration of many young men and women to the German *Reich*, and young men in both villages were called to arms during the war. At the same time, pressures brought to bear on the area during these years served to reinforce the importance of subsistence production and landholding. Military food demands syphoned off much village produce with little remuneration, actually reducing cash income and leaving the villagers more dependent than ever on what they could produce for themselves. In the later war years and the immediate postwar period, there was a general collapse of the Italian market economy. During this time subsistence agriculture in the Upper Anaunia served its citizens well—in favorable contrast to their market-dependent countrymen. Not until the post-World War II "wonder" development of the European economy in general, and of northern Italy in particular, were substantial pressures for change brought to bear on the area.

Following the war, the Upper Anaunia still had nothing new to offer the market, but the market had much to offer it. The first step in commercial involvement was the discovery that a number of things thus far produced locally in inefficient ways could now be bought cheaply in the market. The most important of these commodities were wheat and wool. The per-hectare yield of wheat had always been low, and about once in three years autumn rains waterlogged the grain harvest. Grain produced in wet years was poor in quality, difficult to mill, and yielded inferior flour. The possibility of obtaining high-grade flour at low prices in the market was therefore appealing to the villagers. Increased purchasing of flour, in turn, lead to a reduction in land planted with wheat and in the amount of wheat produced locally. The story of local wool production is similar. The winter feeding of animals and the laborious process of spinning raw wool into yarn made cheap, store-bought, processed wool—or even manufactured clothing—an attractive alternative to home production; sheep-raising thus began to decline soon after 1950. At about the same time restrictions placed on the grazing of sheep in wooded areas, in the interests of conservation, increased the burden of keeping sheep and contributed to the depletion of village herds. In St. Felix, the sheep herd dropped from 250 head in 1948 to 65 in 1953; in Tret the figure dropped during the same period from about 200 to 20.

With less land devoted to wheat, there was a corresponding increase in the amount of land devoted to meadows. Thus, more hay could be harvested to stall-feed animals during the winter. Moreover, with fewer sheep, the hay could be used entirely for cattle. An extra head of cattle, or two, could be added to the herd with no increase in the amount of labor needed to operate the holding and without money investment. Although exact figures showing an absolute rise in Upper Anaunian cattle are not available, the number of cattle sent from St. Felix to the mountain pastures for summer grazing increased from 76 to 109 during the period 1948–1954.

Once under way the conversion of plowland to meadow and increased emphasis on animal husbandry was stepped up rapidly. Since a three-field rotation system had been used to keep fields fertile, a reduction in the amount of wheat planted upset the agricultural pattern. Potatoes (or oats) to wheat to rye was the usual sequence for any given field. When the percentage of wheat to other crops was somewhat reduced, the total number of hectares planted was also reduced in an attempt to retain the three-field system. This simultaneously made it possible to increase cash income: new meadows meant more hay—and another cow in the stall. At the same time, money took on new importance. The growing dependence on cash was clearly demonstrated in 1965 when a poor potato harvest, caused by bad weather, caught a number of families without enough potatoes to fill the need for both food and seed potatoes. Hence, seed potatoes had to be purchased on the market. Previously the amount of land planted to potatoes had been adequate to meet all family needs, even in bad years; in good years, there was a surplus to sell or give away. This was no longer true in the sixties.

While such changes brought an increase in cash income, the source of such income remained the same: the sale of heifers and steers by Felixers, and the sale of calves, butter, and cheese by

Trettners. Most of the increased cash income was used to compensate for the reduction of home-grown produce; the increase in real income was small. However, the continuing emphasis on animal husbandry prepared the villagers for the next step in market involvement—the sale of milk to outside dairies, a step taken in the 1960s. In the case of Tret, the impetus to sell milk came from outside, in the form of an invitation to join a newly established cooperative dairy in Fondo. In St. Felix, it was the result of action taken by village entrepreneurs, who established themselves as middlemen and arranged sales to a commercial dairy.

Throughout the 1960s, the expanding economy of northern Italy, and of Europe at large, affected the local economy in still other ways:

1. As the European economy became more prosperous, the production of fruit in the Etsch and Noce valleys not only recovered, but was intensified, and demands for both seasonal and year-round labor increased. Girls and young men from both communities began to respond to these demands, going to the lowlands in the fall to pick apples and pears, and working in the large cooperatives and privately owned fruit warehouses and processing plants. In addition, the Anaunians found a new craft specialization: the manufacture of wooden fruit boxes. These are made on consignment, or craftsmen simply sell boxes to warehousers who come in search of them in the fall. This manufacture, while relatively unprofitable in terms of returns for the amount of labor and capital expended, can be carried out during the winter months, when other opportunities for earnings are practically nil.

2. Together with the increasing prosperity of Europe, the tourist trade has increased by leaps and bounds. Local young people have found seasonal and full-time employment as chambermaids, cooks, waiters and waitresses, bell hops, and in related kinds of work in hotels, inns, and restaurants. They have found work within the provinces of Bozen and Trento, as well as in Austria and Switzerland. Meran and Lana to the north have became major tourist centers, and during the last decade the communities along the Gampen Road have also found the operation of tourist facilities profitable. Since about 1960, a few minor facilities for lodging and feeding tourists have appeared even in Tret and St. Felix. A number of families have set up rooms and apartments that can be rented during the summer months, and local innkeepers have experienced a small increase in business.

3. With increasing prosperity and rising population in the towns, the building trades have experienced an upswing in business, to the benefit of the local population. Masons from Tret can easily find employment in the lowlands, and a number of them have been able to obtain nearly full-time employment in the surrounding villages. The opportunities were so obvious that many young men from St. Felix began to usurp the traditional prerogatives of Trettners, going to work as stonemasons or as mason's helpers. For a time, lumber brought a good price, and local lumbering activities also were increased. Both the local carpenters and the blacksmith were able to find new kinds of work, the carpenters making doors and windows for new houses, the smith, iron railings for balconies and porches.

4. People from both villages have migrated to the industrial centers of Germany and northern Italy. Partly as a result of this and partly because of the opportunities within the surrounding region, migration to the Americas from St. Felix has stopped entirely. In Tret it continues, but at a greatly reduced rate.

5. An aspect of economic development that is both a part of the development itself and a factor serving to accelerate it, is an increased emphasis on education. Before World War II, villagers attended school for four years. Since the war, mandatory education has been extended to eight years, and opportunities for trade schools beyond this limit abound. These schools are related to both the tourist and building trades, involving training in skills such as cooking, waiting on tables, stonemasonry, laying ceramic tiles, and roofing. The increase in training facilities and opportunities for full or nearly full-time employment in a trade have provided economic alternatives for villagers who wish to stay in the region.

As a result of economic development, land is no longer the only means to material well-being. First, in terms of cash income, the owner-operator of even a large holding in the Upper Anaunia cannot favorably compare with the most menial laborer's income in the lowlands, a fact that is apparent to virtually every villager; even a stonemason's helper earns more cash annually. Second, the opening of a market for milk, with commercial opportunities brought within easy reach of the villages by the Gampen Road, has led estate managers to deemphasize subsistence production and produce for the market. Yet, these changes have not met with the same response in Tret and St. Felix, despite their similar ecological adaptations and the virtually identical impact of market demands on the two villages.

THE RESPONSE IN TRET

In Tret, the alternatives of employment outside the village are extremely popular, to the detriment of the local economy. The village youth *all* seek employment outside of the village after completing their required schooling. Some of them go on to train in a trade or profession; others start work immediately in a job that either requires no special preparation, or provides on-the-job training. The employment of the landless population is no problem, as adequate work opportunities now exist in the lowlands to accommodate all who seek employment there. Instead, the problem now, as seen from the viewpoint of older villagers, is how to get all the work done in the village. Parental attitudes toward the new economic order are therefore ambivalent: the elders are pleased that the present generation does not face the problem of scarce employment they knew in their own youth, but they would like to keep at least one child at home to ease the burden of agricultural work as they grow older. The traditional organization, with its ideology of equal inheritance for all offspring, does not place the responsibility for taking over a holding on any particular individual. Where formerly the managerial position was an enviable status sought by the majority of the population, today it is regarded as a burden. The younger generation is willing to help out with the agricultural tasks at periods of peak activity, principally the harvest, and they spend their Sundays, holidays, and periods of unemployment with their parents; but few of them plan to succeed their parents in the operation of the holding. The young women flatly state they will not marry a peasant, and young men are equally anxious to avoid staying on the land. Only four permanent new households dependent on the local economy were established in the village between 1955 and 1965, as newly married couples set up housekeeping in quarters near their place of employment.

There has arisen a situation of potential conflict both between generations and among siblings. The traditional social organization of the village makes offspring responsible for the care of their parents in old age. Aging village parents continue to interpret this in a traditional way, insisting that at least one of their offspring remain in the village to help with agricultural chores, and to succeed them in operation of the holding. Young people regard remaining at home as a hardship, a sacrifice to their parents. After staying home for any period of time longer than a few months, they invariably put out a call to another sibling to replace them, to give them a chance to earn some money and engage in nonvillage life. If the request is not answered, bitterness grows among the siblings. When none of the offspring responds to the parents' call, tensions mount between the generations.

Young people are not unresponsive to the traditional demand to care for their aging parents. They object only to the traditional *method* of support: they are more than willing to make generous contributions of money, both to parents and to any sibling remaining at home to help them, and to give expensive gifts at every opportunity. In only three cases do sons in their twenties remain with their parents; in nineteen other cases the parents have either accepted the inevitability of finding no heir to the land, or temporary compromise solutions have been made. The young people spend most of their time working away from the village, but they return for a few months at harvest time, or in emergencies (as when a parent becomes too ill to work), make frequent gifts of money and useful or attractive objects, and avoid making a formal commitment to a career outside the village as long as their parents still live (see Table 11).

An even stronger commitment to job opportunities outside the village is characteristic of both middle-aged and younger men with little or no land of their own—the group traditionally forming the fringe population of the village. Without land or parental ties to bind them to the village, they have had no difficulty deciding to seek nonagricultural employment. Except for a few individuals with physical or mental handicaps, the exodus of this group has been complete. They have been joined by a number of the small landholders, men who cannot support themselves entirely from their lands. Some have left the village altogether; others maintain their families in the village and work elsewhere, spending only weekends, holidays, and vacations at home. A few among the small landholders who support their family by working outside of the village still keep up minimum agricultural operations, planting one or two fields and perhaps keeping a cow. Others have rented their land or lent it to a village relative. Although they never constitute a large percentage of the village population, this fringe group in the past performed important functions: they supplied labor at periods of peak agricultural activity, and they were the local craftsmen, and the local entrepreneurs. The loss of this population and the services they performed thus marks a further change in the traditional economy.

Table 11
STATUS OF POTENTIAL HEIRS IN ST. FELIX AND TRET

Age (years)[a]	No heirs[b]	Heir apparent[c]	Minor children	New family	Unmarried siblings
St. Felix Holdings					
>81	—	—	—	—	—
71–80	1	2	—	—	—
61–70	1	9	1	—	—
51–60	1	6	8	—	—
41–50	—	—	8	1	—
31–40	—	—	12	1	1
21–30	—	—	5	2	1
Nónes-Speaking Tret Holdings					
>81	1	—	—	—	—
71–80	5	1	—	—	—
61–70	7	1	1	—	—
51–60	5	1	4	—	—
41–50	—	—	4	1	2
31–40	—	—	2	1	3
21–30	—	—	—	—	—

[a] Age of male manager of the holding. If the holding is under joint management, the age of the youngest comanager is used.

[b] The manager either has no children or adopted heir, or all children have left the village and do not intend to return.

[c] One or more adult children or adopted heirs (20 or more years of age) remain at home and intended to succeed to management of the holding.

Table 12
AGE OF CURRENT ESTATE MANAGERS[a]

Age[b] (years)	St. Felix	Tret	
		Nónes-speaking	German-speaking
	No. (%)	No. (%)	No. (%)
>70	3 (5.0)	7 (17.8)	0 (00.0)
61–70	11 (18.3)	9 (23.0)	1 (16.7)
51–60	15 (25.0)	10 (25.6)	4 (66.7)
41–50	8 (13.3)	7 (18.0)	0 (00.0)
31–40	15 (25.0)	6 (15.4)	1 (16.7)
21–30	8 (13.3)	0 (00.0)	0 (00.0)
Total:	60 (99.9)	39 (99.8)	6 (100.1)

[a] Note that this table refers only to estate managers. There are 15 domestic groups in Tret and 16 in St. Felix that lack estates.

[b] Age of male manager of holding. If the holding is under joint management, the age of the *youngest* comanager has been used.

As a result of the wholesale emigration of the youth and the fringe population, nearly 40% of the holdings in Tret are held and operated by old couples: individuals over sixty years of age, who receive a minimum of assistance in working their holdings (see Table 12). They are too old to keep their holdings as productive as in the past; production on these estates has decreased. This tendency is reinforced by the limited consumption demands of the oldsters. Despite the increasing quantity and variety of material goods available on the market, the old villagers work only to keep themselves fed, and clothed, and to keep in repair the inventory they already possess; they do not seek new commodities. A number of them, now "retired," are doing almost nothing with the land, and derive full support from a combination of small government pensions and contributions from their absent children.

Like others in the area, many of the aging estate managers have also shifted from production for subsistence to the sale of dairy products for the market. However, since their goal is to satisfy decreasing needs with ever decreasing labor, their operations also grow less productive; there has been a relative increase in market over subsistence activities, but a decline in overall production. Whereas the conversion of plowland to meadow has, if anything, proceeded at a greater rate among this group than among others, the increase in meadow land has been used to decrease the amount of work in keeping cattle, rather than to finance a higher standard of living or to modernize their holdings. The yield per unit of land and also per man-hour of labor is considerably higher upon village meadows than upon mountain meadows. Reducing plowland and increasing the size of home meadows means that more fodder can now be obtained for livestock with less effort. This, in turn, has led them to neglect the mountain meadows: some have been sold or rented, and those that have been kept are often neither mowed nor tended. In a few cases, more hay is harvested from the home meadows alone than is needed, and hay is being sold—a situation unheard of under earlier economic conditions.

Manufactured goods, which were not a part of village culture prior to World War II, have appeared in their households, and some agricultural machinery is now used, but none of it has been financed through the operation of their holdings—it has all been donated by offspring working away from the village. (The new items donated to the parents by their children are not always appreciated; a fair number of them quickly disappear from sight.)

Younger villagers who are currently raising families, men between thirty and fifty, with middle-sized and larger holdings, are operating their holdings closer to the potential of the land and are more interested in keeping up production. These households are participating fully in the transition to dairy operations, and manage their holdings in such a way as to keep as many cattle as possible. Nevertheless, the technology of production, including both equipment and techniques of handling animals, has changed little to the present time. The increased cash earnings are partly spent in buying things villagers formerly grew or made for themselves, so that the increase in real income is not as great as the increase in the amount of cash available. Increases in amounts earned are not invested in capital improvements, but are used to increase the consumption of material goods—especially fashionable clothing, home furnishings, and improvements in living quarters. There are, however, three villagers who have gone all out in the conversion to commercial farming, who have added new equipment, such as tractors and mowing machines, and have furnished their barns with automatic watering systems and other improvements. They seek out information on ways to ensure the health and sanitation of their animals, and apply what they have learned. Two of these innovators have borrowed money to make these technological improvements. They cannot, therefore, count their cash returns as pure profit, but must set aside a portion to repay the loans and interest changes.

A half-dozen other villagers, in this age group and intent on staying in the village, are seeking to improve their economic position without commercializing their holdings. They seek, instead, additional cash income by renting rooms or apartments to summer vacationers, by manufacturing wooden fruit boxes, or by taking on an entirely new sideline—one operates a taxi, another has established a bar, and a third is a full-time house painter.

Further evidence that the new economic order is finding favor within Tret is indicated by the movement of a number of owners of medium- and large-sized holdings out of the village. Since 1945, six villagers have sold their entire holdings, one has moved to the city and rented his estate, and two others currently have holdings up for sale. As we have seen, villagers in years past have occasionally sold a field to another villager or to an outsider, but never—except in cases of extreme financial crisis—have they sold an entire holding. Short of outright bankruptcy, with the sale of the holding ordered by a court, sale of an entire holding in the past was very rare. Furthermore, whenever land was put up for sale, whether an entire holding or a field or two, it was usually sold to another villager. Everyone was

"Three villagers have gone all out in the conversion to commercial farming, who have added new equipment, such as tractors and mowing machines."

interested in increasing the size of his holding, and although money has never been plentiful in the village, one or another of the villagers always seemed to have some available. If one had no personal savings, a kinsman (perhaps a "rich" migrant living in the United States) would often loan the necessary amount or buy the land outright, giving the villager the right to buy it from him as money became available. Currently, however, with more money on hand, and with nonvillage kinsmen "richer" than ever before, no villager has been interested in buying the land put up for sale. The house of one of the holders who sold out went to a fellow villager, but all of the other houses and every bit of land were sold to outsiders. The holding that was rented went to a nonvillager, as well. In every case but one of the sales, the purchaser has been a German-speaking Tyrolese, and in the single case of purchase by a family from the Trentino, the land is not being farmed; the new owners operate a sawmill and a general store and have rented out all of the productive land.

THE RESPONSE IN ST. FELIX

Certain aspects of the current economic situation in St. Felix resemble those in Tret. Most village youths find work outside the village soon after they complete their schooling; the emphasis in operation of the holdings is increasingly being directed toward commercial milk production; and the fringe population of the village has been greatly reduced. Money is sent home by unmarried children who work outside the village, and cars, fashionable clothing, and other consumer items are beginning to appear. However, the responses to the pressures of northern Italy's new-found economic prosperity are, in total effect, markedly different from the responses in Tret. In Tret, as we have seen, most villagers are abandoning their peasant identity through the use of strategies that move them off the land into new ways of life. In St. Felix, most strategies aim not to leave the land, but to increase one's chances for success on the land.

Virtually all the young men and women of St. Felix have worked outside the village by the time they are young adults. With the current favorable employment situation men who do not have the prospect of an inheritance and women without marriage prospects to a village landholder are able to establish themselves in a career outside St. Felix with relative ease. There is enough confidence in the stability of these opportunities that these individuals no longer seek to maintain ties with the natal holding as insurance against bad times. They are able to achieve a degree of independence that their landless uncles and granduncles never could have realized in a lifetime. Gifts of labor and cash, once given to keep alive ties to the home holding, rarely are made to brothers who have taken over the family estate; sibling ties are often severed even before the death or incapacity of the father. Contributions of cash and gifts are made to parents by unmarried Felixers working away from the village, but they do not contribute earnings for as many years, nor do they contribute as much as their counterparts in Tret. In addition to young people, older members of the fringe population also have left St. Felix to seek employment. Thus, the dependent, fringe population of the village has been greatly reduced.

However, not all the disinherited youth of the village have cut their ties with the village. Eleven, although they work outside of the village, have been able to secure a small piece of land where they have built a house. Some of them have married and have families.

When possible, they have added a few fields, spending whatever time they can working this land. These part-time cultivators hope to use their earnings to expand their toehold in the village into a full-sized holding, from which they can earn a living without outside employment.

Many young men and women have left the village for outside careers, but the succession to estate ownership and operation is not threatened, as in Tret. At least one sibling in every household plans to assume control of the familial holding. This is true not only of the large- and middle-sized estates, but of small- and dwarf-sized home-steads as well. Many of the younger generation in St. Felix are as anxious to take advantage of the careers available outside the area as their Tret counterparts. There are, however, others in St. Felix who continue to prize the prestigious position of owner-operator, even though the income one can earn from the land may be lower than that from alternate pursuits.

Once continued ownership of a holding is assured, the land-owners aim, as they did in the past, to gain the best possible living from the land. They take full advantage of current market oppor-tunities. Sales of livestock, always a practice in the past, have in-creased during and following the 1950s, but the growing responsive-ness to market demands is evident in the rapid shift recently from raising heifers and steers to milk production.

Cash earnings are employed differently than in the past. Money is now used to buy food once produced on the holding. Other goods formerly produced by the household, and occasionally sold when there was a need for cash, now can be bought cheaply and easily in the marketplace. Some cash earnings are reinvested as "operating capital." Earnings above this amount represent an increase in the real income of a family. Such income is most frequently used to increase the scope of the estate's economic operations; currently this means to expand dairy production. It is possible to improve the quality of cattle already owned or to add to the number of animals, but the critical variable in further expansion is the availability of feed. To produce more feed means to find more land on which to grow fodder and, in the face of a decline in traditionally available labor, the acquisition of more land also requires finding the means to work it.

The search for new meadows within St. Felix quickly exhausted the available supply; ecological limitations make further expansion of meadow land within existing community boundaries impossible. But in Tret, in Fondo, and in other Nónes-speaking villages to the south, landowners are continuing to abandon the land in favor of other economic opportunities. Many fields taken out of production in Tret are thus being taken up by German speakers from St. Felix, either through purchase or rental. In some cases—as in the one outlined above—entire holdings have passed into Tyrolese hands.

Faced with the need for more land and with fewer hands to work it, the St. Felixers are also eager to reduce manual labor through the acquisition of labor-saving machinery: washing machines to free women for work in the fields; mowing machines, suitable for use on village meadows and the more level mountain meadows; tractors, wagons, and power saws, as well as mechanical devices to shred and store hay. In contrast to Tret, where virtually all new machinery is bought with money donated by relatives working outside the village, improvements in St. Felix are being paid for by the landholder through income from agriculture.

In rare cases, a villager will have enough money on hand from past savings to make an outright purchase of land or machinery. More often he will have to borrow money; villagers increasingly seek government assistance in obtaining loans and are willing to use their holdings as security. However, the gross annual cash income of a medium-sized holding, perhaps $1000, may still not cover the cost of a new field or a new machine. After living expenses, taxes, and the costs of operating the holding are subtracted, the $120 yearly payment on a five-year loan of $600 for a hand-operated mowing machine can indeed be difficult to meet. While all villagers would like to own labor-saving equipment, narrow "profit margins" still prevent many from doing so.

Nonetheless, the German speakers are committed to the modernization of agriculture and animal husbandry, and such cash as does become available beyond the daily expenses of the household is invested in the operation. This necessarily means a postponement in the purchase of consumer items and the retention of a living standard that differs little from the traditional one. The contrast in consumption patterns in St. Felix and Tret, where money is more often converted into consumer goods, is strikingly evident in household furnishings. Home furnishings in St. Felix are overwhelmingly homemade or produced by village craftsmen; the house and barn show much construction and repair work of the "do-it-yourself" variety; clothing is still either homemade or, if purchased, is in Tyrolese tradition. Conversely, in Tret, commercially manufactured furniture, curtains, and rugs are much in evidence; repair work and additions to house and barn are done by specialists, often from outside the village; clothing is purchased with an eye toward the styles depicted in national magazines.

In St. Felix small landholders direct production strategies toward supplementing agricultural income with additional employment, as in the past; the more lucrative work opportunities found today outside the village are seen as advantageous because they make it easier to remain on the land. St. Felixers devote a minimum of time to nonagricultural activities. They will work to earn sufficient cash to make up the difference between income from the land and living expenses. They will also work to earn enough cash to pay for a specific capital improvement in their holding, such as a piece of land or a mowing machine. When possible, St. Felixers prefer to remain in the village to work, even for lower wages, and they are quick to exploit any new economic opportunity that arises at home.

The construction of new houses in the village has created one such opportunity, and a small corps of part-time masons and carpenters has formed among the small landholders. In addition, two local entrepreneurs, both small landholders, have been responsible for making contact with dairies, each of them manipulating available credit to buy a truck to carry milk to market, not only from St. Felix, but also from Tret and other nearby villages. Still other villagers have purchased trucks or tractor-trailers and engage in both local and long-distance transport. Trucks have replaced the horse-drawn wagons that used to carry lumber out of the area, and several men with tractor-trailers now do much of the local hauling—manure from barn to field, hay from field to stall, and other odd jobs.

TRET AND ST. FELIX COMPARED

The economic development in northern Italy and neighboring countries since the 1950s has produced a situation in which both villages can successfully direct the exploitation of holdings to growing market demand. At the same time, villagers now can opt for careers in the surrounding regions, careers that produce material rewards equal or better than the land. In both St. Felix and Tret, estate managers and owners whose social and economic predominance was once assured through the exercise of restricted rights of access to land, no longer can draw upon a labor pool of the disinherited and disadvantaged. On the contrary, the security of present-day careers outside the villages has released the surplus population from dependence on the family estate, and has virtually eliminated the fringe population.

Despite these similarities, St. Felix and Tret have responded in markedly different ways to economic developments. In Tret, roughly 16% of the landowners have abandoned the land for careers outside the village; the remainder barely holds its own on the land. Young people are overwhelmingly choosing careers outside agriculture, renouncing tradition for urban patterns of consumption. In other words, the village increasingly exports labor and skills, receiving, in return, material support for the marginal and the old.

St. Felixers have also given up a past emphasis on subsistence production in favor of production for market, but use this link to the market to expand and intensify a continuing commitment to life on the land. Even a number of those who, in the past, would have been counted among the disinherited, and denied any opportunity for an independent living in the village, are now using the alternatives offered by economic expansion to finance a return to the land. To a certain degree, the expansion and intensification of agriculture and livestock-keeping in St. Felix is made possible by the abandonment of agriculture and herding by their neighbors in Tret and other Nónes-speaking villages to the south.

THE MATERIAL BASIS OF THE GOOD LIFE

Change is in the air in the Anaunia; yet for most families in St. Felix and Tret, control of the resources that make up an estate remains a major goal in their lives. Estate managers and owners are involved in new commitments to the market, but their present economic potential is still strongly determined by the productive capacity of the estate. Villagers still take the measure of a man by how closely he approaches the potential of his holding. They know that men differ in ability and motivations, that some are more skilled than others in handling livestock and in coaxing the land to yield a maximum harvest—and they continue to assign prestige rating in these terms.

At the same time, villagers know that every holding presents its own problems. Land will vary in productivity, depending on the quality of soil, amount of moisture, degree of slope, altitude, and exposure to sunlight. When villagers estimate the capacity of a holding to yield a living, they understand that many such factors vitiate an exact correlation between estate size and the productivity of a holding. When they evaluate property, it is not in terms of numbers of hectares, but in terms of potential harvests.

One useful rule of thumb used for assessing the productive potential of estates in the past was to estimate the number of cattle that could be supported. This estimate then was combined with an appraisal of the estate's minimal capacity to produce crops and, in St. Felix, with a judgment on the size and quality of the holding's stand of timber. Today, when holdings are increasingly dependent on the sale of milk, herd size is a more exact measure of estate value. Three or four head of cattle represent, to a villager, the smallest number needed to provide an acceptable living. Five to eight head of cattle will provide an income that allows full satisfaction of cultural expectations. Men with nine head of cattle are regarded as well-to-do (see Table 13).

Under present conditions, however, families who combine a smaller number of cattle with a craft or trade can command an income equal to that produced by one of the more profitable estates. In the past, small- or dwarf-sized estates could do no more than meet minimum expectations, even when cultivation and livestock-keeping were combined with a sideline. Today, owners of such estates respond more readily than others to opportunities presented by a growing market; with rising incomes, some of them have been more successful than neighbors with more productive estates.

In the past, a landless wage laborer could not maintain a household in St. Felix or Tret. Today several households are headed by men in this category (see Table 13, footnotes b and c). Most dependent on the national market for food, clothing, and furnishings, they are also the most attuned among villagers to national views of a reasonable standard of living. Their influence, together with that of others who now live and work in the lowlands, but continue to maintain ties with kinsmen in St. Felix and Tret, is beginning to reshape the cultural expectations of villagers. This influence is probably more pronounced in Tret than in St. Felix, but in both villages, there is a growing awareness of how local life styles compare with national norms.

Still, the new economic order within the villages remains firmly rooted in a history of subsistence production. Earlier economic conditions continue to exert an influence: they still affect the composition of estates and the skills of current village residents. How well villagers are able to maintain an acceptable life style depends in

No. of cattle	St. Felix		Tret	
	No. of households[a]	Economic sidelines[b]	No. of households[c]	Economic sidelines[b]
A. Holdings "well off"				
9–12	7	7 (no sideline)		
7–8	1	1 (mayor of St. Felix)	1	1 (no sideline)
2–3			1	1 (blacksmith, trucking)
1–2	1	1 (general contractor)		
			1	1 (sawmill)
	Total: 9 (14.7%)		Total: 3 (6.4%)	
B. Holdings meeting village expectations				
5–8	14	11 (no sideline) 1 (rake-maker, church sexton) 1 (inn, general store) 1 (forester)	4	3 (no sideline) 1 (herdsman)
3–4	2	2 (no sideline)	10	5 (no sideline) 1 (inn) 1 (donations from absent children) 1 (cabinet-maker) 1 (inn, butcher) 1 (stonemason)
			3	1 (general store) 1 (inn, commercial poultry- and rabbit-raising) 1 (worker in Switzerland)
	Total: 16 (26.2%)		Total: 17 (36.2%)	
C. Holdings meeting minimum village expectations				
3–4	17	15 (no sideline) 1 (sawmill, grain mill) 1 (carpenter)	3	3 (no sideline)
1–2	7	3 (stonemason) 1 (inn, general store)	5	2 (stonemason) 1 (donations from absent children)

(continued)

Table 13 (continued)

No. of cattle	St. Felix		Tret	
	No. of households[a]	Economic sidelines[b]	No. of households[c]	Economic sidelines[b]
		1 (inn, bakery)		1 (taxi)
		1 (rake-maker)		1 cabinet-maker)
		1 (trucking)		1 (inn, donations from absent children)
			4	1 (stonemason)
				1 (laborer in Canada)
				1 (pension, donations from absent children)
				1 (electrician)
	Total: 24 (39.4%)		Total: 12 (25.5%)	

D. *Holdings falling below village expectations*

No. of cattle	St. Felix		Tret	
1–2	9	4 (no sideline)	8	8 (no sideline)
		3 (rake-maker)		
		1 (inn)		
	3	2 (no sideline)	7	4 (no sideline)
		1 (pension)		3 (pension)
	Total: 12 (19.7%)		Total: 15 (31.9%)	
	Total: 61 (100.0%)		**Total:** 47 (100.0%)	

[a] In addition to the 61 households in St. Felix that derive at least some income from an estate, there are 9 other independent households that hold no land at all. Three derive their income locally: one as a mail carrier, one as an employee of the commune, and the third as a carpenter. Three others work outside the Upper Anaunia; the final 3 are pensioned workers. Four apartments are maintained by men who work and live outside the Upper Anaunia, but who hope to buy an estate someday and take up full-time residence in the village. There is also an apartment for the village priest and his housekeeper.

[b] Additional income for all households in Categories C and D is derived from wage labor either within the Upper Anaunia or the surrounding lowlands. In addition, some households derive income from the practice of minor crafts such as making fruit boxes, from occasional donations from absent siblings or children, or from the infrequent practice of a major craft, such as carpentry or stonemasonry.

[c] In addition to the 47 households in Tret that derive some portion of their income from an estate, there are 7 other independent households that hold no land at all. One receives a state salary for delivering mail. Three others are supported by men who work outside the Upper Anaunia; the final 3 are families that formerly earned their living outside the village, but returned to it in retirement. Six apartments are maintained by villagers who now live and work elsewhere. Three of these own estates, but have rented all their land, while the other 4 own no land at all. All use these apartments occasionally on weekends, holidays, and during the heat of summer.

large measure upon the correspondence between their abilities and resources and present-day opportunities.

PROSPERITY AND MISFORTUNE

Let us illustrate the range of outcomes in the struggle for a living by describing a successful household and one that operates on the edge of misfortune.

Among the ingredients necessary to a satisfactory existence in the Upper Anaunia is wine. Wine is the preferred drink throughout the Trentino–South Tyrol, and since no grapes grow in the upper reaches of the valley, villagers must procure it from the outside. The local stores carry wine in bulk, as well as bottled wine, so that cost-conscious villagers can bring their own bottles to be filled. Many find this an acceptable arrangement, but others prefer to keep a quantity of wine in their own cellars, either because they wish to be self-reliant or because this way is least expensive. These villagers may buy grapes in the fall and make their own wine, or else fill their casks at one of the lowland wineries. Much of the wine thus obtained comes from the vicinity of Kaltern, lying just over the Mendel Pass to the east.

The village blacksmith in Tret is one of those who finds it convient to keep a good supply of wine on hand: he has casks sufficient to hold over 900 liters of wine in his cellar. Once a year the casks are carried to the lowlands in his truck and filled. Household requirements are such that within ten months, the supply is nearly exhausted; a supplemental run will be made to fill a few small casks to tide the family over until the main run in the fall. The smith himself has several glasses of wine at each meal (breakfast excepted), and, as is customary in this part of the world, he takes a glass to satisfy his thirst throughout the day. In addition, his son, although he drinks less than his father, consumes his daily share, and the women of the household, too, find an occasional glass of wine refreshing.

It is in the nature of a smith's work that customers call frequently at his shop with horses, oxen, or cattle to be shod; metal tools or wagons to be repaired; or with a request that the smith come to them to perform some task of maintenance or repair. Furthermore, it has become fashionable to enclose porches and outside staircases with iron, rather than wooden, rails; the smith must be visited in

"The village blacksmith in Tret keeps a good supply of wine on hand."

order to arrange their manufacture. While an animal may be shod without sharing a glass of wine with the smith, any larger undertaking involves some discussion, which is most easily carried out over a few glasses of wine—no contract would seem valid unless sealed with a drink. In any case, it is pleasant to pause for a moment over a glass of wine to gossip about village affairs. Thus, the 900 liters are consumed in ten months.

In years past the smith was one of those who bought his wine in Kaltern. Several years ago, while returning from a Sunday visit to a daughter who is a cook at a resort on Lake Garda, he chanced to stop and sample a glass of wine in the village of Nomi. He pronounced this wine to be the best he had ever tasted. The *cantina* that produced the wine was located, and thereafter an annual pilgrimage has been made by truck to fill the smith's casks with Nomi wine. The trip has become an event in the smith's household, developing a routine of its own. The empty casks are loaded on the truck the night before the trip. Since the smith does not drive, he and his son undertake the two-hour drive in the morning, with the son at the wheel and the smith at his side. At the *cantina,* a cooperative effort is made by father and son: the father samples and passes judgment on the wine, and the son oversees the filling of the casks. The *cantina* personnel are most generous with samples of wine, the smith most appreciative of their generosity. There is inevitably some delay before the cask filling begins. A tacit understanding exists that the son will demonstrate less appreciation for the *cantina's* generosity, in the interest of getting the truck and its cargo successfully home. Finally the casks are filled, and the trip home made, with a stop along the way for lunch.

Once home, the smith, glass in hand, exercises a loose supervision over the unloading, while the delicate task of actually getting the barrels of wine safely off the truck and into the cellar is performed by the son, who is assisted by several friends. Wine is, of course, offered to all helpers, and, in fact, to anyone whom the smith observes within hailing distance. Good will and contentment prevail, and the smith is expansive: it is a time to reflect on the blessings of life.

At this season of the year, the harvest is either in or, barring some utterly unforeseen catastrope, safely predictable. The hay has all been cut and stowed in the barn: the harvest has been good and the hay is adequate to feed the smith's two cattle through the winter and into spring. Milk from the cattle is now being sold to a commercial dairy near Cles, furnishing the family with a welcome supplement to its income from the smithy. In addition, there are two calves, both born four months earlier, which will be sold to the butcher in two more months. Since May two pigs have been in the stall, as well as some fifteen chickens. The swine will be slaughtered soon after the new year and processed into *Bauernspeck,* the main meat staple of the Tyrolese peasant, and sausage. The thought of the slabs of meat soon to be hanging in his smokehouse gives the smith almost as much pleasure as the full casks of wine.

The potato harvest this year has not matched expectations. Why this should be so is puzzling to the smith: a good field and good seed potatoes were used, the field was well manured, and rainfall has been plentiful. But, no matter. The harvest is still adequate: there are enough potatoes stored in the cellar to last the family until the next harvest and to feed to the pigs as well. It was planned to buy new seed potatoes next year anyway. No wheat or rye was planted in the smith's fields this year, nor has there been for a number of years: bread is available, delivered fresh each day to the village stores from a bakery in Fondo. It is not expensive, and so the smith's family, like most, no longer bakes bread.

No cabbages have been planted either. The family of the son's fiancée has a large cabbage patch and has promised to give the smith as many as needed. Soon the cabbages will be harvested, and the women will make the winter's supply of sauerkraut. While all families rely on sauerkraut for their winter vegetable, most switch to fresh vegetables in the spring. Here the smith is unique: he has little use for fresh vegetables, preferring sauerkraut at all times of the year.

The government forester, a neighbor and friend of the smith, has made the annual allotment of timber to the families in Tret. The smith's family has already gathered in the wood, separating lumber for building repairs, and beginning the sawing and chopping of firewood into stove-sized pieces. That the forester is a friend to the smith, many would claim, is not irrelevant to the matter of the timber allotment: while each year a specific amount, 9 cubic meters is, in 1965, due every family in Tret, the forester actually makes the allotment by assigning specific trees to each family. Since the assignments are made on the basis of standing (or felled) whole trees, rather than of sawed and measured wood, there is considerable variation in the amount of timber the various families actually receive. The forester insists that the assignments are made on a purely random basis— others are not so sure.

The smith anticipates no hardships in the foreseeable future. All the family is healthy, and the essentials needed to keep family and livestock throughout the winter are on hand. Furthermore, there is enough work lined up to occupy both the smith and his son for several months—more than they can get done, and the smith knows even more is forthcoming. The monthly income produced by the smithy during the winter will be little less than in the summer—here the smith feels he is more fortunate than most of his neighbors, who must rely more on the land or on a combination of land and work such as stonemasonry. The winter provides these families with little opportunity to add to the annual income.

In recent years the smith has found many new outlets for his talents, which have enabled him to increase his cash income and, hence, to rely more on trade and less on the land. At the same time, the nature of his work has changed. The demand for iron railings and the increasing mechanization of agriculture have meant that the smith uses his welding iron more frequently than his hammer. His workshop now resembles a machine shop as much as it does a smithy.

In his contemplation of the growing income from his trade, the smith's reflections are paralleled by hopes held by many fellow villagers for the sale of milk to lowland dairies and the opportunities for work as laborers, at ever increasing wages. A lessening of the toils of mountain agriculture seems near, as cash income increases and peasants can purchase farm machinery. Still, in his winter stores, the smith reflects the wisdom of a tradition of subsistence production. This is shared by his fellow villagers, who appreciate the ability of the land to provide them directly with the necessities of life.

The prevailing attitude within the villages is one of guarded optimism toward present and future. But not everyone is optimistic. Even villagers who feel that their material circumstances have improved measurably in recent years maintain a cyclical, rather than a progressive, view of economic trends. Their fathers have seen both good times and bad on the land, and while times are now good, bad will follow, just as surely as night follows day. A time will come again when jobs cannot be found outside the valley, and everyone will return to subsistence production.

For other villagers, the realities of the present weigh heavily, and contemplation of the future produces only despair. This is the case with the small landholder in St. Felix, Johann A. His dwarf holding lies at the lower end of the altitude range used by villagers for meadow and plowland, but it is excessively stony, with numerous outcroppings of bedrock, and has no convenient source of water. The estate was established around the turn of the century by Johann's father, the eighth of eleven siblings. Born on a medium-sized estate, his father enjoyed no prospects of success in the village, but managed to accumulate a small savings by working outside St. Felix. With this money he purchased the bit of land now worked by his son. This land had belonged to another village estate, which, through the centuries, had gathered firewood and grazed animals there, but had never thought it worth the effort to clear. Johann's father cleared about 5000 square meters for plowland and meadow, and just after 1900 he built a small house–barn complex on one edge of the clearing. His holding was enlarged slightly by a few parcels of land inherited by his wife.

Johann married and took over management of the holding a few years before his father's death in 1960. At about the same time, he was also able to purchase a mountain meadow of about one-half hectare. Still the holding could not support his family, which in 1966 included six children as well as an ancient, widowed mother. In addition to working the land, therefore, Johann must continually seek other sources of income. Thus, he finds himself in the classic circumstance of the manager of a dwarf estate in the uplands.

Just as the smith, in an ebullient mood, summed up his life situation for us on a pleasant, fall day, so Johann, on the edge of despair, related his plight one rainy day in the summer of 1966, as we sat at his kitchen table.

With insufficient plowland, he can grow only garden vegetables and potatoes, not rye or wheat. Therefore, he has to buy all his bread and has to do without straw for his animals. His meadows, too, are insufficient, even for his single milk cow and two calves, so he must buy additional hay every year. Consequently, his cow does not receive adequate feed and gives less milk than she should—not even enough for his children—so he also must buy milk daily. He has not been able to afford a wagon, and, in any case, he could not afford to keep a draft animal or hire one from a neighbor. Instead, he transports hay, potatoes, and wood from field to home in a wheelbarrow. Thus, tasks take him longer than anyone else.

With a holding too small to provide his family with even a subsistence living, Johann A. is unable to join fellow villagers who sell milk for cash. Instead, he must find ways of earning money to keep his estate operating and to feed his family. Unfortunately, he possesses no special skills marketable in the modern world. In the past, he made and sold rakes, worked as a herdsman, and hired out as a laborer in an attempt to make ends meet, but of late he has given up rake-making because it is too time consuming, and he has not been offered any herding jobs for some time. He has thus been depending on occasional employment to supplement his inadequate income from the land. This income has been too low over the past several years.

In 1965, Johann sold the half hectare of mountain meadow in order to keep his family fed. By the following summer, this money was spent, and he hoped to go to Austria after the harvest to work as a lumberjack. During the summer, however, an old bachelor uncle who had been living alone was taken mortally ill and was brought to Johann's home. Although completely bedridden, the old man clung tenaciously to life.

To make absolutely certain that we understood the hopelessness of his situation, Johann took us to the bedroom where the old man lay. Once there, he explained that the old man refused to die.

Worse still, he was helpless, but had the appetite of a healthy man. Because of him, Johann could not go to Austria as he had planned. Johann's wife could not possibly care for six children and a dying uncle-in-law, carry on daily household chores, and assume his winter chores of caring for the animals and cutting firewood. Johann did not know how he would manage to earn the additional income he so desperately required, but he would have to remain the winter in St. Felix. As we left the bedroom and he closed the door behind us, he gazed sadly at the old man and said, "Er *will* nicht sterben"— "He will not die." Not only fate, but a stubborn old man who refused to die when his time had come, conspired against him.

Fate had reserved two more blows for Johann that year. When working in the mountains, helping another villager get in his hay, a fully loaded hay wagon fell on him. Broken bones and internal injuries kept him hospitalized for several months. What Johann's wife needed to keep the family alive she obtained on credit from local stores and from neighbors. As we prepared to leave the village, he had returned home and was struggling to regain his health. His family's mounting debts were causing him severe anguish, and he had no idea how he could repay. He worried, too, about his wife, who now had an invalid husband to care for, as well as a dying old man. In the meantime, the uncle still breathed, but the youngest child did not survive a severe cold that had developed into pneumonia.

If the smith's well-stocked cellar reflects the wisdom of a tradition of subsistence agriculture, Johann's plight demonstrates its drawbacks. Nearing fifty years of age and tied to a small holding insufficient to meet his needs, Johann has been unable to make ends meet and has no reserve from which to meet the inevitable emergencies a family faces from time to time. Moreover, with all his energies directed toward day-to-day survival, he cannot explore the alternatives that have become available in recent years. He is trapped by the economic circumstances of the past.

CHAPTER **X**

Kith and Kin

In Chapter VIII we noted that St. Felix and Tret converge in their ecologically grounded practices of inheritance, but diverge in their inheritance ideologies. We then suggested that the ideology of inheritance, in each case, could not be understood as long as one thought of it as a mere cognitive "mirror image" of actual practice; it had to be seen in reference to the way the inheritance process produced public outcomes. The roles of heirs and the disinherited, of managers and coheirs, are played out in the public as well as the domestic sphere; they affect family relations and juro-political relations. We must therefore seek to understand the relations that structure the public domain, the ways in which villagers in St. Felix and Tret manage kinship relations of consanguinity and affinity and ties with kith—friends and neighbors. We will note again that the two villages converge in social practices that resolve common ecological problems, but diverge in the social ends to which their patterns of behavior are directed. We shall find that they resemble each other in the outward form of domestic groupings and in the ways in which affines are recruited into kin groups. However, St. Felix and Tret differ in the ways authority is exercised and obeyed, as well as in the ways domestic groups relate to each other through cooperation or competition.

DOMESTIC GROUPS

If we look first at the social demography of the two villages, we note that they tend to be much alike in the distribution of types of domestic groups (see Table 14). According to the Italian census of 1961, Tret had a population of 238; our data show this population to be divided among 60 domestic groups. St. Felix at that time had a population of 335; our data show a total of 76 domestic groups for St. Felix. When we analyze the internal composition of these units, we find the distribution of alternative types given in Table 14.

In Tret, we find 34 domestic groups, or 58.6%, of the total consisting of conjugal families, and 7 domestic groups, or 11.66%, consisting of conjugal families plus some member of the parental generation or sibling set of the conjugal partners. Thirteen domestic

Table 14
TYPES OF DOMESTIC GROUPS IN ST. FELIX AND TRET

Domestic type	St. Felix	Tret
Conjugal pair	8	9
Conjugal pair, children	43	25
Conjugal pair, children, members of the parental generation	4	4
Conjugal pair, children, siblings of either parent	2	3
Sibling set: unmarried brothers and sisters	2	6
Sibling set: unmarried brothers and sisters, other kin	1	—
Married woman, children, husband dead or absent	6	5
Married woman, children, husband absent, husband's sister	1	—
Married man, children, wife absent	—	1
Married man, operating singly	4	—
Married woman, operating singly	—	5
Unmarried man, operating singly	3	2
Unmarried woman, operating singly	1	—
Priest and housekeeper	1	—
Total:	76	60

groups, or 21.6%—more than one-fifth—include only one adult of either sex.

When we look at the data for St. Felix, we immediately notice the preponderance of conjugal families *with offspring*. Domestic groups consisting of conjugal families alone number 51, or 67.1%, of the total; however, whereas in Tret every third conjugal family has no children, in St. Felix this is true only of about every sixth family. The percentage of domestic groups housing joint or extended family units is somewhat smaller than in Tret: only 7.9%. The number and percentage of groups containing only one adult of either sex is somewhat comparable: there are in St. Felix 14 such units, or 19.7%, of all domestic groups. In Tret they number 13, constituting 21.7% of all domestic groups. This is also true of the groups consisting of sibling sets or jointly managed by adults connected through kinship: 6 in Tret, 4 in St. Felix. Striking, however, is the presence of married women managing their affairs alone in Tret, and their absence in St. Felix; conversely, there are 4 married men running their own domestic unit in St. Felix, none in Tret. Again, the priest and his housekeeper form an integral part of the community in St. Felix; in Tret, the priest resides in Fondo and only visits the village on weekends and holy days.

These figures, however, are deceptive, for they offer only a static picture, a snapshot taken at one point in time; they reveal nothing of the ongoing process of recruitment and loss through which domestic groups continuously change composition. We have seen that the inhabitants of both villages compete with each other for available resources and employment opportunities; not every one can stay at home and hope to raise a family. We have also seen that land provides the prerequisite for a stable livelihood, as well as for the few social honors available within the communities: but land is limited, and only a few villagers can succeed to ownership or management. The rest must take on additional jobs or seek work beyond the village confines.

Membership in a domestic unit is gained in several ways, though most depend on kinship: one enters a domestic group by birth, or by marriage to a member. Occasionally, individuals gain membership through adoption, as when a childless couple or a set of unmarried siblings adopt a nephew as heir to the estate. On occasion, membership may also be acquired through contract, as when a "hired man" takes up residence in return for room or board. In all cases, membership is validated by residence (both full-time and occasional) and by participation in the family's social and economic life.

All domestic units form economic entities, but the individuals who compose them stand in variable relations to the operations of that entity. Some individuals form the core of the domestic group. These are the men and women who reside in the village, work the same land, maintain a herd in common, and share in the yield of agriculture or livestock-keeping. If they are occasionally employed outside the estate, or if they practice a craft, their earnings are generally pooled with those of other core members for their joint support and welfare. At the same time, a domestic group will also include fringe members who manage their resources separately and are often away from the holding or the village, but who contribute both money and labor more or less regularly to the unit. In return, they can count on periodic support from the estate.

The variable distribution of individuals who form a domestic group with regard to its management and operations is evident at any one time, but it is itself both function and outcome of processes of differentiation that operate over time. To understand domestic groups in St. Felix and Tret, we must include the vital dimension of change. The domestic group may indeed be associated with a particular set of resources, an association that often persists over generations; its membership, and the social relations among its members, are forever in flux. Members are born, marry, emigrate, and die. Within each set of siblings, some will eventually succeed to the management of the holding, marry, and raise children; others remain single and subordinate to the estate manager. Still other kin will drift from full dependence on the natal holding into fringe relationships, and may in the end become fully divorced from its operations. Later, as the manager grows older, there will be increasing pressure on him to retire in favor of an heir. Each domestic group thus goes through a cycle, differing in composition and internal social arrangements at each stage in its development.

The three phases in the developmental cycle of the domestic group conceptualized by Meyer Fortes (1958) illuminate these processes as they take place in St. Felix and Tret. The two communities exhibit similar phases of expansion, dispersion, and replacement in the development of their domestic groups. In this similarity lies one of the reasons for convergence in the patterns of their social organization.

The first phase, that of expansion, begins with the formation of a husband–wife dyad, to which children are normally added. The domestic group at this stage may include, in addition to the conjugal family, one or more siblings of either marriage partner, or one or more dependent parents. Throughout this phase, children are fully dependent on their parents. During childhood, their contribution to the operation of the unit is minimal, amounting certainly to less than the cost of their support. But as they grow into their teens, their ability to work becomes a significant resource. Children contribute to the operation of the estate, and their labor is used by neighbors in labor exchanges, or in return for a small wage. Periodically they also leave their homes to work in adjacent villages or in the lowlands: this is true especially at harvest time. Where and when children will be permitted or required to work is up to the father, and their earnings will go into the family coffers. As they grow older, some children—though not all—spend more time working away from home than working at home, and return to the village only occasionally.

The second phase is that of dispersion. It begins when the first of the siblings cuts himself adrift from the family estate and asserts his independence by making his own decisions about matters that affect his life. Although this move occasionally is quite abrupt, and the protagonist involved is "here today and gone tomorrow," it usually involves a gradual process. The young man or woman simply begins to decide for himself when he will work away from home; and instead of turning over all earnings automatically to the father, he will now manage his own earnings, though continuing to donate both money and labor to the natal estate. This places the individual in a fringe relation to the holding. His resources no longer are pooled with those of the estate, but any donations of cash and labor create a fund of obligation on the holding on which he will be able to draw during periods of unemployment or illness.

Many individuals will remain in such a fringe relationship for the rest of their lives; others, however, will ultimately sever ties completely, either through permanent emigration, through marriage to a villager already in control of his or her own land, or through unusual luck in locating an independent source of livelihood. These villagers retain neither rights in, nor obligations to, the natal holding. Similarly, any sibling in a fringe relationship who no longer makes donations to the domestic group thereby loses any claim upon its resources. A move from dependence to complete independence, therefore, always involves a decision on the part of the individual as to whether his best chances lie in eventual succession to management of a holding, or whether they are favored by opportunities available to him outside the village.

The third phase, that of replacement, is initiated by the retirement or death of the landowning parent, and hence by the assumption of decision-making by the filial generation. The domestic group can now assume two alternative forms. In the first, the holding will continue to be managed by the sibling subset resident on the holding, and functioning as a unit. This set, with perhaps one or more dependent parents and some siblings on the fringe, will now constitute the domestic group. In about 10% of all cases, it will persist in this form, none of the members ever marrying and bringing a spouse onto the estate.

The second alternative is more common: one of the brothers eventually marries and establishes himself as manager of the estate. His dominance is clear and it will pass to his offspring, even if some of his siblings remain on the holding. Still, he will have to contend with legitimate claims to support from the holding by retired parents or any fringe siblings. The resolution of these claims is again a matter of time: eventually the parents will die; his own wife and their children will crowd out the manager's siblings; resident siblings will move into a fringe relationship; and those on the fringe will become independent.

Thus, the cycle returns to the initial stage. It should be noted that the estate rarely goes without an heir in the succeeding generation when a holding is operated by a sibling subset. One of the sisters may have an illegitimate child who can take over the holding, or a child may be adopted. The adopted heir apparent will preferably be the son of a sibling who has married into another village holding. This nephew will have no prospects of inheritance because numerous brothers and sisters are rivals for any claim he might have to the estate.

It is obvious that domestic units in the villages do not have synchronized developmental cycles, and that at any point in time not all domestic groups show the same composition. While at a given period there seem to be alternative *kinds* of domestic units (nuclear families, extended families, and sibling sets), these simply represent a single type of organization at different *stages* in the developmental cycle. One domestic group will, in all likelihood, pass through all of these forms in the lifetime of each generation.

During all stages of the developmental cycle, core domestic units usually constitute a single household, but, under certain circumstances, they may contain two or (rarely) more households. When the heir marries and begins a family before either of his parents has died, each nuclear family may occupy a separate apartment, taking meals and keeping house separately, but sharing dependence on the same resources. Separate households are even more likely when two or more of the heir's siblings are core members of the unit, especially when at least one of them is female.

In the Upper Anaunia, mothers, sisters, and wives play basically interchangeable roles in the functioning of the domestic unit. In the event of the death or incapacity of the mother, a daughter is expected to take her place as domestic manager. As long as the holding is operated by her father or by a sibling subset, the daughter will serve as mistress of the house and ruler of the kitchen. However, as the Tyrolese say, there can only be one woman in a kitchen; thus, when a brother marries and brings home his bride, domestic responsibilities are reassigned among the women. The new bride will, of course, expect to be mistress of the house, and in this claim she will be supported by her husband; but the sister who has occupied this position can only regard her as an intruder. If the sister is too firmly committed to a life on the holding to leave, the situation can be resolved only by the establishment of a second dwelling—the new wife caring for her husband, and the sister caring for the other unmarried siblings resident on the holding. Although two households are created, they remain part of a single domestic unit. The land that supports both remains undivided and will be worked jointly under the direction of the married heir; the second household will dissolve only upon the death of its members.

Each phase thus defines who may and may not be a member of the domestic group, and what degree of membership he or she holds at any given time. Yet each phase also involves succession to the pivotal managerial position, through which the resources of the household are marshalled and distributed to members, and denied nonmembers. While the requirement that the resources of a household be handled by a single manager is common to both villages, they diverge sharply in the structural and cultural definitions of the managerial role and in its consequences for social relations. In St. Felix, managerial authority is yielded to a single heir, to the exclusion of all other heirs; in Tret decision-making *authority* is vested in one sibling, while other siblings keep an active say in decision-making. This operates synchronically to *exclude* collaterals and affines from influencing decision-making in St. Felix, and to *include* collaterals and also affines in Tret: it also operates diachronically, to structure out exclusive lines of descent in St. Felix, and in contrast, to favor the creation of wide-ranging inclusive networks of kin and affines in Tret.

These differences are reflected in the status terminology in current use in the two villages. Status terminology may refer to kinship statuses or to statuses held by virtue of other criteria. The strategic status designations in St. Felix are the nonkin statuses of *pauer* and *päuerin* (*Bauer, Bäuerin*), denoting at one and the same time the role of male head of household: the individual who makes independent decisions regarding the management of the household and of the land. Which *pauer* you are talking about is specified by adding the name of the homestead. If the homestead is called *Prunn* (*Brunn*), the *pauer* on *Prunn* is the *Prunnerpauer*, his wife the *Prunnerpäuerin*, and his unmarried children the *Prunnerhans* and the *Prunnerlísl*. The family occupying a particular homestead is thus identified with the homestead; frequently people in St. Felix are hard put to remember a family name, such as Geiser or Kofler. Possession of a homestead at any given time creates a lineal link with past or future holders of that homestead, even though family names on an estate may change through inheritance, sale, or transmission to an affine. Thus, Jakob Geiser is the *Prunnerpauer* until he transmits the holding to his son Joseph; Joseph then becomes the *Prunnerpauer* and Jakob retires to the position of being the *Prunnerfater* (*Brunnervater*). But Jakob Geiser may sell his holding to Franz Bertagnolli, who becomes the *Prunnerpauer* in his stead. Jakob may leave only a female heir; the man she marries—Joseph Kofler—becomes the new *Prunnerpauer*. The occupation of a homestead in St. Felix, and especially occupation of one of the "historic" homesteads created before the beginning of the seventeenth century, thus creates an ongoing set of public statuses that may be held in lineal fashion whether or not there exists actual lineality in kinship affiliation.

Kinship terminology in St. Felix offers further indication that the emphasis, in reckoning kinship as in reckoning the occupancy of a holding, is on lineal continuity. Note in the diagram that follows the clear distinction between lineals and collaterals:

	guck-nen = Gr Gr Fa	
	guck-nándl = Gr Gr Mo	
	nen, opa = Gr Fa	
	nena, nándl, oma = Gr Mo	
fetter = Fa Br	*tata, fater* = Fa	*fetter* = Mo Br
basl = Fa Si	*mama, muoter* = Mo	*basl* = Mo Si
fetter = cousin	*ego*	*fetter* = cousin
basl = cousin	*bruoder* = Br	*basl* = cousin
fetterle = Br So	*swester* = Si	*fetterle* = Si So
basl = Br Da	*bua* = son	*basl* = Si Da
	madele = da	
	enkel = Gr So	
	enkelin = Gr Da	

While lineal relatives are specified, siblings of the father and mother, as well as their offspring, are lumped into terms that serve to distinguish collaterals (and their claims) from the family line—thought to have exclusive rights against all other family lines.

In Tret, lineal restrictiveness is absent. Here, too, homesteads have names, but the concept of the corporate homestead enduring over generations is absent. Also lacking are the designations of houses—so common in St. Felix—by names pertaining to the cultural landscape, such as the Felixer *Prunn*, spring, *Pruck*, bridge, and *Roan*, cultivated field; names deriving from events, *Erspam*, initial construction; or names referring to the occupation of an early occupant, for example, *Binder*, cooper. Instead, the names of houses in Tret indicate descent from an apical and genealogically verifiable ancestor: houses occupied by the descendants of a Floriano are identified as Floriani; houses occupied by the descendants of a Venanzio as Venanzioti. They may be called by the personal nickname of some past occupant, such as *topa*, mole, or *kappa*, horseshoe. Emphasis here is not on exclusive, perpetual lineality, but rather on continuity of membership in a patrilineally related stock. Where St. Felix emphasizes the exclusive dominion of a *pauer* on his homestead, at the expense of any collaterals, Tret indicates that the community is made up of Venanzioti, Rossi, Floriani, Cru, Franzelini, and Bassoni. Kinship terminology used in Tret distinguishes collaterals from lineals, as in St. Felix; however, the tendency to specify generation in collaterality, rather than to lump kinsmen into a general cousin category is consonant with the social fact that ties between lineals and collaterals are maintained and ever woven anew. Tret's terminology is as follows:

non = Gr Fa
nona = Gr Mo

barba = Fa Br	*pare* = Fa	*barba* = Mo Br
anda = Fa Si	*mare* = Mo	*anda* = Mo Si
cozín = Fa Br So, Fa Si So	*ego*	*cozín* = Mo Br So, Mo Si So
cozína = Fa Br Da, Fa Si Da	*fradel* = Br	*cozína* = Mo Br Da, Mo Si Da
	sorela = Si	
neú = Br So, Si So	*fiol* = So	*neú* = Br So, Si So
nésa = Br Da, Si Da	*fioli* = Da	*nésa* = Br Da, Si Da

nipo = Gr So
nipota = Gr Da

AUTHORITY

The divergence in status terminology—both kin and nonkin—between St. Felix and Tret points to more thoroughgoing structural distinctions in social organization. Nonkin status terminology in St. Felix emphasizes the importance of the holder of a homestead and all other terms are derivative from it; kinship terminology similarly underlines lineal continuity, rather than collateral spread. Terminology thus mirrors what is evident in social relationships: the dominant social relation in St. Felix is that of the publicly recognized role-set of homesteader and his heir. Implicit in this usage is an ideological view of the juro-political field: it points to a conceptualization of St. Felix as a sum total of independent domains, each held by an independent operator who, in turn, passes on his undivided and exclusive jurisdiction to one and only one male descendant in the filial generation. Further, implicit in this usage is the idea that landholders and heirs of the "traditional" domains held before the beginning of the eighteenth century take social precedence over landholders and heirs of holdings founded after that date. At the same time, the publicly recognized role of homestead holder is isomorphic, within the context of the family, with the kinship role of father, the male head of household to whom wife and children are subservient. He is the *pauer*, the sole authority in managing the family resources.

We have seen that the eldest son is raised from birth to be the successor of his father, and we have noted the conflicts generated between father and son who compete for authority within the household; one must yield if the other is to come into his own. We have also seen that frequently this silent battle leads to the departure of the intended heir: the eldest, reared to compete with the patriarch, cannot stand his subservient role and leaves the holding; the next oldest sibling then takes his place.

The battle for authority is not only fought between father and heir apparent, it is also fought between older and younger brothers, and among all brothers and all sisters. Brothers are fierce competitors for inheritance; because only one can inherit, the others must either depart or resign themselves to serve on their brother's estate as *knecht*, as family servant in return for maintenance. A man can be either *pauer* or *knecht*, but not both: if he is the first, all authority is his; if the second, he has no autonomy in decision-making and is likely to remain unmarried, a *bua* (son) forever. Brothers also compete with sisters, for a sister will either marry and contend for family resources to obtain a dowry, or she will remain an unmarried dependent, entitled to support in times of illness and need, but subject to her brother's authority.

The fierce competition for dominance in the family has further consequences. The *pauer*, sole heir to authority, remains alone with his family upon the holding; all siblings who do not accept the stipulated conditions of dependence must leave the homestead. If they remain in the village, relations will tend to be cool, especially in cases where the new heir does not wish to recognize any claims by a sibling whom he has relegated to the status of the "undeserving poor." If they leave the village, relations will become minimal. If they go abroad, they may not be heard of again, and should an estate be intestate, the priest—charged with finding the migrant siblings—may be unable to locate any of them. If relations with collaterals in St. Felix are cool, so are relations with affines.

There is hardly any visiting from house to house in St. Felix. Men meet at the tavern, in church, and at council meetings. Women meet at the store and at church; but the warm friendliness of food and wine shared between friends and neighbors in the household kitchen, so characteristic of Tret, hardly exists in St. Felix. Transactions between households, such as exchanges of labor or favors, are carefully tallied and precisely reciprocated (see Chapter VII).

The Felixers are indeed unable to "tolerate one-way flows [Sahlins 1965b: 148]." They not only make certain that all such flows are immediately reciprocated; there is a positive avoidance of situations that could result in such transactions and that would raise the specter of either the donor's or the recipient's inability to respond with appropriate repayment. Each *pauer* in St. Felix is, in Curate Hillebrand's words, "an emperor on his own homestead." Oscar Lewis (1970), studying families in Tepoztlán, has called this kind of family "monolithic"—one in which all authority is concentrated in the hands of the father, and all other relations within the family, including the relation of husband and wife, are subservient to his power. In St. Felix, the dominance of the monolithic family inhibits the development of freely flowing reciprocal relationships between the members of the various domestic groups. The only time this exclusiveness is diminished is when a prospective son-in-law freely offers labor to his future in-laws. During such periods prospective brothers-in-law often develop bonds of common sympathy, but these grow weak after the marriage has taken place.

Tret faces a similar problem of resource management; here, too, a population confronts scarce resources in its search for sustenance.

However, whereas the monolithic family is characteristic of St. Felix, Tret approximates the type Oscar Lewis called "segmental." Such a family is characterized by differentiation and segmentation in structure, and it consists of "clearly delineated interaction units or subgroups." He points out that in this type of family, each parent maintains a distinguishable and separate mode of interaction with each of the children, and he sees the mother as playing" a crucial role in the communication system, acting as an intermediary between husband and children, and between her sons and daughters [Lewis 1970: 280]."

In Tret, authority is not vested exclusively in the male head of household; wife and husband complement each other in its exercise, and each participates in a distinct subset of relations with the daughters and sons of the household. Underlying this division and balancing of authority may well be the expectation of partibility in inheritance, which allows both husband and wife to be potential claimants to property and which divides the estate ideally among the several siblings. The Tret family may be compared to a severalty, a shareholding corporation, while the Felixers cleave rigidly to the concept of each man a lord on his domain. The severalty always produces a manager; but even when the other siblings leave the holding and community, they always retain some ideal share of access, some claim upon the resources of their kinsmen.

Marriage outside the village is as frequent in Tret as in St. Felix; in both villages there is a tendency to slough off women through marriage and through the export of unmarried females. The sociological consequences of this outmigration are different, however, because emigrants from Tret remain in extended communication with their kinsmen at home: relations and gift giving not only persist, but affinal as well as collateral ties are cherished and remembered. Hence, they are capable of social use in times of need. A couple and their in-laws are not mutually antagonistic occupants of different homesteads. A son-in-law (zénder) has easy and free-floating relations with his father-in-law (misiér, madón) and mother-in-law (madóna); and with their sons, brothers-in-law (cuña), and daughters, sisters-in-law (cuñada) he also maintains an easy relationship. In Tret, collateral and affinal kinsmen, as well as neighbors, can drawn upon each other's services, and these ties frequently crosscut generations. Thus, a man may call on his mother's sister's sons to help him cut hay, or two men who share a common great-grandfather may maintain solidary relations in forming the nucleus of the Socialist Party in Tret. Collaterality, affinity, and easy rela-

"In Tret, collateral and affinal kinsmen, as well as neighbors, can draw upon each other's services."

tions with a neighbor (*auzín*) are reinforced by continuous visiting and sustained by a pattern of generalized reciprocity, in which services offered are not immediately tallied and reciprocated. It is notable that the German-speaking families who have migrated into Tret during the second quarter of the twentieth century and taken up agricultural holdings there soon relinquished patterns of balanced reciprocity in favor of the generalized reciprocal patterns characteristic of their Romance-speaking neighbors.

We return to our basic thesis that, to the extent that both villages respond to similar ecological influences, the basic forms of their social organization tend to be much alike. Both face the problem of inclusion and exclusion of members, and both face the problem of resource management. While the functions of management are located in a single pivotal role, it is in the way this role is played out that the two communities diverge most. They differ in the definition of authority and in the ideological component of the authoritative role; this difference has maximal importance for the structuring of the social field—into exclusive lineages of homesteaders in St. Felix, and into the creation of an open and interlaced network of relations in Tret.

MARRIAGE

We have seen that domestic units in St. Felix and Tret converge in their formal composition, but that they differ in the ways social relations are played out both within and between such units. Marriage has a special role in mediating social relations. In both villages, the importance of marriage is underlined through public recognition of the married state: only a married person fits the social definition of an adult male or female. There are some bachelors and spinsters, and it is possible for them to set up independent households. However, such arrangements are felt to be lacking in social gravity; a single person needs to demonstrate some rare skill or possess unusual wealth to compensate for this handicapped participation in public life. Only marriage is thought to impart to a person the full range of social responsibility, and the single person remains in the social category of the unmarried young, who fail to accept full responsibility. In Tret, a male remains a *put*, a boy, rather than becoming an *om*, a man; the single girl, a *puta*, rather than becoming a *femna*, a woman. In St. Felix he remains a *bua*; she a *madl*. Neither a *put* nor a *bua* can be a full player in the social games;

Table

MARRIAGE AND FERTILITY AMONG THE MOST NUMEROUS

Period	All marriages	No. childless marriages	Childless marriages (%)	Total no. of children born	No. of children per marriage	No. of children per child-bearing marriage
						St. Felix: Geiser,
1700–49	13	6	46.1	39	3	5.57
1750–99	13	4	30.7	66	5.07	7.35
1800–49	35	4	11.4	219	6	7.06
1850–99	32	3	9.3	183	5.7	6.3
1900–49	23	8	34.7	86	3.73	5.73
						Tret: Bertagnolli,
1800–49	25	1	4	129	6.16	5.37
1850–99	53	5	9.4	327	6.18	6.81
1900–49	51	17	33.3	166	3.25	4.9

lacking children, they do not build for the next generation. In St. Felix, this means that an unmarried person cannot ensure the continuity of the homestead; in Tret, the single person—bereft of affines—remains an unreliable and imperfect participant in social life.

In both villages, a wedding signifies public recognition that a new domestic group is about to be formed; the formation of a new domestic group means that one signals publicly one's willingness to take part in the village's collective life. A married couple becomes a point of social anchorage in the ongoing flow of village life. Yet, because social relations flow so differently within and between domestic units in the two villages, we should also expect differences to show themselves in the patterns established through marriage. To understand how marriage operates in Tret and St. Felix, we shall first examine it in its ecological context, then probe for its sociological consequences.

Marriage has at least two important ecological aspects. First, it is a mechanism for the social production of children (Terray 1969: 155–156): in each generation, marriage produces the next generation of dependents on resources. Second, it is a mechanism for differentiating within each generation those who are given privileged anchorage in the village through marriage from those who are only granted diminished anchorage or denied roots altogether. This is

15
NAME CATEGORIES: ST. FELIX AND TRET

Total no. children surviving past age 15	Children surviving past 15 (% of total)	No. M surviving past 15	No. F surviving past 15	Ratio M to F survivors	No. of surviving children per marriage	No. of surviving children per child-bearing marriage
Kofler, and Weiss						
32	82	14	18	0.78/1	2.46	4.57
47	71.2	24	23	1.04/1	3.6	5.2
174	79.4	83	91	0.91/1	4.97	5.6
139	75.9	75	64	1.71/1	4.34	4.79
75	87.2	39	36	1.08/1	3.26	5
Donà, and Profaizer						
123	95.3	80	43	1.86/1	4.9	5.12
263	80.45	132	131	1/1	4.9	5.4
140	85.8	75	65	1.15/1	2.7	4.1

a point perhaps not sufficiently stressed in the anthropological literature, perhaps because attention has been focused on marriage as a mechanism for recruiting *additional* personnel to the local group, to the neglect of viewing marriage in its negative aspect—as a mechanism for ridding the villages of surplus contenders for local resources. The anthropological concern with processes of recruitment through marriage, in turn, seems to stem from an overly localized focus of study, which ignores the wider ecological and social contexts within which local groups functions. Hence, it seems useful to emphasize that marriage can also operate in the opposite direction, by redistributing possible consumers of resources among other localities, or by denying marriage to certain members of the population.

To enable us to see trends in the social production of offspring in Tret and St. Felix, we have analyzed the extant data on marriages and fertility for the most numerous name categories in both villages. These are the Geiser, Kofler, and Weiss in St. Felix, and the Bertagnolli, Donà, and Profaizer in Tret (see Table 15). Disregarding the data for the eighteenth century as we lack reliable comparable information for Tret, we can note certain parallel phenomena in the two villages during the course of the nineteenth century and the first half of the twentieth. With some surprise, we discovered that the number of children surviving past age 15 per childbearing couple did not diminish significantly over that period. The number reached a high in St. Felix during the first half of the nineteenth century (5.6) and a low in Tret during the first half of the twentieth (4.1). In general, however, it has oscillated around the number 5.0 in both communities during this century and a half.

What has significantly changed, however, is the percentage of married couples who have remained childless. Consisting of approximately 10% of all couples in the nineteenth century, the number increased to 33% in the twentieth century—to 34.7% in St. Felix, to 33.3% in Tret. This shift effectively reduced the number of potential exploiters of resources in the villages during the first half of the twentieth century. This phenomenon follows a very high rate of survival of children during the second half of the nineteenth century, the period of greatest population build-up in both villages. While the average number of children surviving past the age of 15 per married couple in St. Felix was 4.79, the lowest figure since 1750, the percentage of childbearing marriages was 90.7, the highest for any half-century. The comparable figures for Tret for this period of population crisis were 5.4 surviving children per childbearing couple, the highest for any half-century since 1800; the percentage of child-bearing couples was 90.6, as compared with 66.7% in the half-century following.

In assessing the role of marriage in population movement, we must visualize the formation of families against the background of changing relations between population and resources. The nineteenth century was in general a period of population build-up, proceeding slowly during the first half of the century, after the introduction of the potato, and reaching the proportions of a crisis during the second half. The effects of growing population pressure on resources is manifest in statistics on emigration. To illustrate the ratio between residents and migrants for any fifty-year period since 1750, we have used data on 74 sibling sets drawn from the largest name category in Tret, the Bertagnolli, and 76 sibling sets of Geisers and Koflers the largest name categories in St. Felix (see Table 16). More than one-third of all male Bertagnolli siblings (or 38%) left Tret during the first half of the century, and nearly half (49%) left during the second. Pressures on the women to seek their fortune elsewhere were even stronger: 64% of all female Bertagnolli siblings left Tret between 1800 and 1849, and 65.9% left between 1850 and 1899, a total close to two-thirds of all Bertagnolli sisters.

Trends in St. Felix are comparable: 34.7%, or more than one-third, of all male Kofler and Geiser siblings left St. Felix between 1800 and 1849; 45.3% left during the second half of the century. The Kofler and Geiser sisters who left amounted to 46% of all sisters, or nearly half, between 1800 and 1849; 88.7% (more than two-thirds) left between 1850 and 1899 (see Table 17).

Table 16
SIBLING SET STUDIES

	St. Felix: Geiser and Kofler		Tret: Bertagnolli	
Year	Sets subjected to marriage analysis	Average size of set[a] (range)	Sets subjected to marriage analysis	Average size of set[a] (range)
1750–99	9	6.8 (3–12)	4	8.5 (7–11)
1800–49	23	7.3 (2–15)	17	5.3 (2–8)
1850–99	25	5.8 (2–15)	33	7.1 (2–16)
1900–49	19	5.6 (2–13)	20	5.5 (2–10)
	Total: 76		Total: 74	

[a] This figure is based on siblings surviving past age 15.

Table 17

	St. Felix: Geiser and Kofler				Tret: Bertagnolli			
	M (%)		F (%)		M (%)		F (%)	
Year	Stay	Leave	Stay	Leave	Stay	Leave	Stay	Leave
Percentages of siblings staying in or leaving the natal village								
1800–49	65.3	34.7	54.0	46.0	62.0	38.0	36.0	64.0
1850–99	54.7	45.3	31.3	68.7	51.0	49.0	34.1	65.9
1900–49	75.6	24.4	61.0	39.0	68.9	31.1	39.2	60.8

	M		F		M		F	
Year	Married	Single	Married	Single	Married	Single	Married	Single
Numbers of siblings staying in the natal village								
1800–49	32	15	13	19	27	11	6	5
1850–99	23	9	10	6	39	13	19	11
1900–49	27	9	15	10	16	9	12	6

Table
SIBLING

Birth of sibling sets (years)	St. Felix: Geiser and Kofler					
	Total		No. who marry		% who marry	
	M	F	M	F	M	F
Marriage						
1800–49	72	58	37	20	51.0	34.0
1850–99	64	48	32	23	50.0	45.8
1900–49	49	38	34	20	70.8	41.7

	M		F	
	Married	Single	Married	Single
Siblings who remain				
1800–49	32 (86.5%)	15 (42.9%)	13 (72.0%)	19 (50.0%)
1850–99	23 (72.0%)	9 (28.1%)	10 (43.0%)	6 (24.0%)
1900–49	27 (79.5%)	9 (60.0%)	15 (80.0%)	10 (55.6%)
Siblings who leave				
1800–49	5 (13.5%)	20 (57.1%)	5 (28.0%)	19 (50.0%)
1850–99	9 (28.0%)	23 (71.9%)	13 (57.0%)	19 (76.0%)
1900–49	7 (20.5%)	6 (40.0%)	5 (20.0%)	8 (44.4%)

What was the role of marriage during this century of population build-up? (See Table 18.) During this period, roughly half the male siblings in both St. Felix and Tret remained celibate; this was also true of two-fifths of the female siblings in Tret and three-fifths in St. Felix. Two-thirds of these unmarried people left the villages to seek their livelihood elsewhere. In other words, an unmarried individual was twice as likely to leave the village as to stay. What, in contrast, was the likelihood of remaining at home for villagers who married? Four out of five brothers, in either St. Felix or Tret, who married stayed in the village; a married male had a four-to-one chance of remaining in his home community; an unmarried male had only one chance in three. A married female in St. Felix had a three-to-one chance of staying during the first half of the century; in the second half, she had fewer than one chance in two. In Tret, there were pressures on both married and unmarried women to leave: between 1800 and 1849, 57% of all Bertagnolli sisters who married, and 71% of those who did not, left Tret. Between 1850

18
STATISTICS: ST. FELIX AND TRET

		Tret: Bertagnolli			
Total		No. who marry		% who marry	
M	F	M	F	M	F
data					
61	31	30	14	49.0	45.0
104	91	48	60	46.6	66.6
45	46	25	32	54.3	69.5

M		F	
Married	Single	Married	Single
in the natal village			
27 (90.0%)	11 (35.5%)	6 (43.0%)	5 (29.4%)
39 (78.4%)	13 (23.2%)	19 (32.0%)	11 (35.5%)
16 (64.0%)	9 (45.0%)	12 (37.5%)	6 (42.8%)
the natal village			
3 (10.0%)	20 (64.5%)	8 (57.0%)	12 (70.6%)
8 (21.6%)	43 (76.8%)	41 (68.0%)	20 (64.5%)
9 (36.0%)	11 (55.0%)	20 (62.5%)	8 (57.2%)

and 1899, the percentage of married Bertagnolli sisters who left Tret had risen to 68%, or more than two-thirds. The comparable figure for unmarried sisters who left was 64.5%. During the nineteenth century, Tret thus favored the emigration of married women substantially more than did St. Felix. Even then, Tret was casting a wider social net, building more active ties with affines, than its German-speaking neighbor.

In the twentieth century the population crisis eased. The population of St. Felix fell from 348 in 1869, to 257 inhabitants in 1930; Tret, in 1930, numbered 237 people (Bertagnolli 1930: 60–61). There was substantial emigration: at least 10% of the population had moved permanently to North America by 1900. The continuous exodus of females, notably in Tret, must have severely diminished the childbearing potential of the population. We have seen that during the nineteenth century, more than two-thirds of all Bertagnolli sisters left Tret to take up residence elsewhere. Moreover, many young men died in World War I, especially men from St. Felix. We have also seen that the percentage of childless families increased markedly in both communities. Surprisingly, the median age of marriage for women did not rise, though men in both communities tend to marry older than in the nineteenth century (see Table 19).

Thinking in purely local terms of the relation of population to resources, one would expect emigration to diminish in the twentieth century. This is true in St. Felix but not in Tret (see Table 17). Tret continues, as in the past century, to export about half its unmarried siblings, both male and female, and three-fifths of its married female siblings. But while in the past only one married male in five moved away from Tret to his wife's village, now only one in three does so. Attachment of males to Tret through marriage is clearly weakening (see Table 20).

St. Felix, on the other hand, has moved toward markedly stronger anchorage for both males and females. The proportion of married to unmarried siblings remains the same as in the past; in the nineteenth century, three-fifths of Geiser and Kofler siblings who remained unmarried had to leave the village permanently, but for the period 1900–1949 this held true for only one-third of the males and one-half the females. On the other hand, 80% of all marrying female siblings and 79.5% of all marrying male siblings now stay in St. Felix, an increase especially in the case of the females. This trend toward increased anchorage and decreased pressures for migration is evident also in the increase of marriages in St. Felix between 1950 and 1969: St. Felix males celebrated forty-eight new marriages compared with only twelve in Tret. In other words, from 1950 to 1969 there were twenty-five marriages per decade, as compared with an average of twenty marriages per decade in the period 1800–1949. There has also been a decrease in the median age of marriage for both males and females in St. Felix during this period (see Table 19); St. Felix has consequently seen the growth of numerous young families with small children.

A variety of factors appear to be at work in creating the contrast between St. Felix and Tret. First, St. Felix clearly has chosen to emphasize agricultural adaptation. Its homesteads appear to be more viable, only 20% (versus 30% in Tret) falling below the minimum of cultural expectations. On the one hand, inheritance ideology has operated to retard the rate of land fragmentation; on the other, permanent or temporary, secondary occupations now can be combined with a commitment to agriculture, to permit a more successful adaptation to the mountain environment. In Tret, interest in cultivation is waning, and industrial or tertiary occupations have kindled the imagination of the younger people to the point that permanent work elsewhere, or emigration, seems increasingly desirable. In contrast to St. Felix, which remains committed to agricultural pursuits, Tret is rapidly moving away from involvement with mountain agriculture.

External political conditions reinforce internal trends. The inclusion of the Upper Anaunia in the Trentino and Italy creates a wider occupational area in which Trettners can exploit their widespread collateral and affinal ties, and take advantage of varied opportunities available. In contrast, access to secondary and tertiary occupations within Italy is restricted for the Tyrolese, either because available jobs are not offered them, or because they do not want to accept employment alongside Italians. Thus, fewer than 3% of the employees in the industrial zone of Bozen are Tyrolese (Leidlmair 1958: 254). The narrower range of jobs combines with an inability to mobilize affinal and collateral ties, to channel the Felixers into what they know best—agriculture. In the first half of the twentieth century and increasingly since World War II, St. Felix has witnessed a rapid increase in childbearing families, most of whom remain tied to the land. On the other hand, the capacity of Tret to retain its population appears to be diminishing in proportion to the pull from commerce and industry in the world beyond the village.

Yet it is also true that some Felixers, like the Trettners, have adopted secondary and tertiary occupations, in addition to their involvement in agriculture and livestock-keeping. We explored these

Table 19
MEDIAN AGE AT MARRIAGE

Years:	1700–49	1750–99	1800–49	1850–99	1900–49	1950–61
Males:						
St. Felix	33	34	31	31	34	28
Tret	—	28	26	30	38	31
Females:						
St. Felix	30	31	27	30	28	26
Tret	—	29	23	24	24	25

new options in Chapter IX. From one viewpoint, this means that the continued, successful adaptation of the German-speaking Tyrolese to the environment through cultivation and the keeping of domesticated animals now is substantially supplemented by income from extralocal sources. One can even say that this monetary supplement has provided the extra margin required to reinforce the traditional roots in the land, and that this extralocal income now sustains the

Table
AGE AT MARRIAGE IN

Age (years)		St. Felix								
		1750–99		1800–49		1850–99		1900–49		1950–69	
		No.	(%)	No.	(%)	No.	(%)	No.	(%)	No.	(%)
	FEMALES									
<20		1	(2.45)	4	(4.39)	0	—	3	(3.30)	2	(4.10)
20–29		18	(43.90)	55	(60.43)	44	(49.40)	49	(52.20)	34	(69.40)
30–39		18	(43.90)	23	(25.21)	32	(36.00)	35	(38.00)	13	(26.50)
40–49		3	(7.30)	9	(9.90)	11	(12.40)	4	(4.30)	0	—
>49		1	(2.45)	0	—	2	(2.20)	1	(1.10)	0	—
	Total:	41	(100)	91	(100)	89	(100)	92	(100)	49	(100)
	MALES									
<20		1	(1.80)	1	(1.00)	0	—	0	—	0	—
20–29		17	(30.80)	34	(36.60)	26	(27.90)	29	(30.90)	26	(54.20)
30–39		23	(41.80)	44	(47.00)	45	(48.40)	38	(40.40)	18	(37.50)
40–49		11	(20.00)	9	(9.70)	17	(18.30)	20	(21.30)	3	(16.20)
>49		3	(5.50)	5	(5.40)	5	(5.40)	7	(7.40)	1	(2.10)
	Total:	55	(100)	93	(100)	93	(100)	94	(100)	48	(100)

Felixers' continued ethnic identity as German Speakers in an Italian State.

The shift to nonsubsistence-oriented activities in St. Felix has produced a significant measure of social change. It has served to narrow the social gulf between wealthier cultivators and poorer peasants, as it has also narrowed the gulf between heirs and the disinherited. To be an heir to one of the "historic" estates still bestows on a villager major social honors. It is now possible, however, for a villager who inherits no land to make a living through nonagricultural pursuits. Thus, disinherited siblings are no longer as completely dependent on the family estate as they were in the past; many more than could before World War II now can stay in St. Felix and hope to marry. At the same time, relations between collaterals remain fraught with tensions: there are still conflicts over authority and inheritance, and even under the best of circumstances, these relations are marked by cool neutrality, rather than active cooperation. Similarly, relations with affines continue to be weak, because such ties, are still economically and socially nonfunctional.

Today in St. Felix the dominance of the exclusive lineal unit is

20
ST. FELIX AND TRET: 1750–1969

	Tret								
1750–99		1800–49		1850–99		1900–49		1950–69	
No. (%)		No. (%)		No. (%)		No. (%)		No. (%)	
FEMALES									
0	—	5	(12.50)	7	(8.75)	7	(5.30)	1	(8.30)
4	(50.00)	26	(65.00)	60	(75.00)	93	(69.90)	10	(83.40)
3	(40.00)	6	(15.00)	11	(13.75)	28	(21.00)	1	(8.30)
1	(10.00)	3	(7.50)	2	(2.50)	4	(3.00)	0	—
0	—	0	—	0	—	1	(0.80)	0	—
8	(100)	40	(100)	80	(100)	133	(100)	12	(100)
MALES									
0	—	1	(1.80)	0	—	0	—	0	—
6	(66.70)	31	(60.80)	39	(46.60)	47	(37.00)	5	(41.70)
3	(33.30)	16	(31.40)	34	(40.50)	64	(50.40)	6	(50.00)
0	—	0	—	9	(10.70)	9	(7.10)	1	(8.30)
0	—	3	(5.90)	2	(2.40)	7	(5.50)	0	—
9	(100)	51	(100)	84	(100)	127	(100)	12	(100)

being challenged, but there has been no accompanying shift to widen the extent of the social network as in Tret. In Tret marriage outside the village and emigration immediately open a network of new relations upon which all kinsmen may capitalize. In St. Felix, the tendency is to remain tied to the village; the emphasis in Tret is on increased involvement with a range of kinsmen over a wide area. Since villagers in St. Felix do not maximize links with affines, they minimize interpersonal relations as a means of connecting themselves to the world beyond. Since villagers in Tret enter into active social interchange with numerous relatives elsewhere, they also maximize channels of communication through which flow information about life in the city. In this sense, St. Felix remains parochial, whereas Tret's social network is extended throughout the wider world.

Felixers thus discourage wider social alliances that might be created through marriage, while Trettners seek to expand the scope of familial relations. We shall do well to keep these structural points in mind as we examine the way each village recruits its spouses. Table 21 tells us something of the marriage ranges of men and women in both Tret and St. Felix. It shows us the degree to which spouses, during the nineteenth century, were drawn either from the home locality or from adjacent localities of the same language category. It also shows us that there was little intermarriage across ethnic boundaries. In the nineteenth century, 85.8% of all St. Felix men married women from St. Felix or Unser Frau; 80.1% of all St. Felix women also married locally. Thus, in the nineteenth century, 80% of all spouses came from the historic *Deutschgegend* of the Upper Anaunia.

In the same century, Tret men married Romance speakers from Tret or from other *frazioni* of the Fondo commune in 73% of the cases, and Tret women in 82.6% of all marriages. This picture has been somewhat modified in the twentieth century: St. Felix men still marry within the *Deutschgegend* in 74.5% of all cases, and St. Felix women in 60.5%. Tret men continue to marry within Fondo in 67.1%, and in the Tret women in 50%, of all marriages. In the twentieth century, there has been a tendency—stronger in Tret than in St. Felix—to expand the marriage range beyond local boundaries.

This expansion beyond the Upper Anaunia has meant that the Felixers now, to an increasing degree, draw spouses from other parts of the German-speaking South Tyrol. Trettners, for their part, find spouses increasingly in other parts of the Trentino, and even in the *Regno,* or "Old Italy," the area of the Italian kingdom before

Table 21
IDENTIFICATION OF MARRIAGE PARTNERS

Origin of spouse	1800–99		1900–61	
	No.	% of total	No.	% of total
Marriages of St. Felix men				
St. Felix	90	60.8	54	55.0
Unser Frau	37	25.0	19	19.5
German-speaking Tyrol	19	13.0	22	22.5
Trentino	2	1.2	3	3.0
Total:	148	100.0	98	100.0
Marriages of St. Felix women				
St. Felix	90	71.4	54	43.5
Unser Frau	11	8.7	21	17.0
German-speaking Tyrol	20	15.9	29	23.4
Trentino	5	4.0	13	10.5
"Old Italy"	—	—	5	4.0
Other	—	—	2	1.6
Total:	126	100.0	124	100.0
Marriages of Tret men				
Tret	60	49.0	38	56.7
Fondo, Dovena, Castelfondo, Malosco, Ruffré	29	24.0	7	10.4
Other parts of Anaunia	30	24.0	11	16.4
Other parts of Trentino	1	1.0	2	3.0
German-speaking Tyrol	3	2.0	8	12.0
Other	—	—	1	1.5
Total:	123	100	67	100.0
Marriages of Tret women				
Tret	60	58.3	38	35.2
Fondo and vicinity	25	24.3	16	14.8
Other parts of Anaunia	11	10.7	18	16.7
Others parts of Trentino	4	3.9	8	7.4
German-speaking Tyrol, mostly immigrants to Tret	2	1.9	20	16.5
"Old Italy"	1	0.9	8	7.4
Total:	103	100.0	108	100.0

World War I. We may interpret this widening of the marriage range as a function of the widening range of occupational contacts, which create social contacts beyond the Anaunia.

At the same time, there has been an increase in interethnic marriages. These, however, bear a special aspect. First, St. Felix men continue to reject marriages with Romance speakers. St. Felix women, on the other hand, in the course of the twentieth century, have married Romance speakers, either from the Trentino or from the *Regno,* in as many as 14.5% of all cases. Tret men show a similar pattern: 12% have married women from the German-speaking South Tyrol. Tret women enter into the highest percentage of interethnic marriages—18.5% of all marriages fall into this category.

To explain differences in interethnic marriage patterns, we must refer to our discussion of relevant structural factors. Quite apart from considerations of ethnic or political identity, marriages of St. Felix men with Romance-speaking women are discouraged by the demands placed upon such marriages by the structures of family and authority in the Tyrolese village. It would be difficult for a Romance-speaking girl, accustomed to more distributive decision-making and to active social relations with collaterals and affines, to adapt herself to the patterns of unitary and exclusive authority characteristic of St. Felix. Conversely, the St. Felix *pauer* would not find in a Romance girl the qualities required of a *päuerin,* as partner in managing his little "imperial domain."

The St. Felix homestead requires of the woman in charge a greater inventory of domestic skills, organized within a much more hierarchical social setting. Therefore, St. Felix women will find it relatively easy to pass from a hierarchical type of family organization to participation in a more segmented and diffuse type of decision-making unit. St. Felix women are aware of this; they say that Italians are more likely to be nice to their women and children than Felixers. Moreover, marriage with Romance speakers brings them into a much wider social network than would marriage in St. Felix, and through this network they are also connected with the world beyond the Anaunia. For many German-speaking women, marriage with Romance speakers can thus mean entry into both a less rigidly structured family and the world of modern consumption and fashion.

Tret men, in turn, welcome marriages with German-speaking women. They value the fact that their wives have been raised in more rigid households: they say that they work harder than Nónes women, and add jokingly that because their hands are larger, they can fashion more sizable dumplings. Most marriages of Tret women with German speakers have not occurred with men from St. Felix, but with immigrants from other German-speaking areas of the South Tyrol—especially from Graun in the Vintschgau—who settled in Tret after World War II. In Tret, these immigrant men did not establish the homestead and authority complex of their Tyrolese home communities. They participate fully in the social life of their Romance-speaking neighbors, adopting also their generalized patterns of reciprocity. The second generation of these Tyrolese immigrants is becoming completely Italianized. In fact, marriages of Trettners with members of this new generation are structurally equivalent to marriages with Romance speakers from Fondo or beyond.

Most interethnic unions, however, do not involve partners from St. Felix and Tret, but German or Romance speakers who come from outside the Upper Anaunia. In this sense, they are comparable to marriages contracted by Felixers outside the *Deutschgegend*, or by Trettners in the *Regno*; they do not cross the boundary line between adjacent communities. Marriage between Felixers and Trettners is rare (see Appendix 4). Of 560 marriages recorded in the marriage register of St. Felix since 1723, 44 involved Romance-speaking partners. Nineteen of these, however, were contracted by persons with only temporary and tangential relations with the village: for example, the nephew of an Italian-speaking priest married a woman from his home village, and two residents of Fondo sought to be married elsewhere than in their home town. Thus, 541 marriages were recorded for villagers proper, only 25 of which were between Felixers and Romance speakers. We also have records of 366 marriages in Tret; 22 of these were between Trettners and German speakers. Of 907 recorded marriages for both Trettners and Felixers, only 20, or 2.2%, were between villagers from Tret and St. Felix. We thus conclude that interethnic marriage is easier when spouses come from communities far apart, when the precise details of each other's background are not known. Because Felixers and Trettners know each other well, they are unlikely to marry.

ASSOCIATIONS

We have noted that domestic groups in Tret and St. Felix differ in the ways they reckon continuity over time, distribute decision-making authority, emphasize or deemphasize ties of collaterality and affinity, and utilize patterns of reciprocity. They differ not only in patterns of kinship and kithship, but also in the numbers and kinds of associations, and in the intensity of organized associational life.

A few associational forms are shared by the two villages. In both St. Felix and Tret informal cliques of men hunt together and celebrate, on rare occasions, success in the hunt. Both villages maintain volunteer squads of fire fighters who not only fight fires, but, clad in special uniforms, form honor guards for the priest during religious processions. In both villages, choral groups regularly sing during church services and annually perform in a secular context. Yet there are more associations in St. Felix, and they are stronger there than in Tret.

St. Felix has a community council, capable of making autonomous decisions with regard to issues that affect the village; Tret does not. Tret is a *frazione*, a dependency of Fondo. Two representatives from Tret are sent to the town council in Fondo, but they have only two votes in a council of twelve, and are subject to veto by the other ten members, who are from the more populous nucleus of Fondo and the *frazione* of Vasio. In addition, much of the routine work of administration in Fondo is performed by the communal secretary who, according to the requirements of the Italian state, is an appointee of the Province and responsible to the provincial government in Trento. Similarly, the forest resources of Tret are managed by a forester appointed by the forest service of Fondo.

St. Felix, in contrast, follows the Tyrolese pattern of rural self-management. Its council of twenty men, headed by a mayor, is elected by the community; the communal secretary is an employee of the council and not of the Province. The council can devote itself to local improvements under its own autonomous charter. It has constructed a communal water delivery system, and constructs and repairs communal roads. It regulates the operation of communal pasture lands and forest, and has recently begun to upgrade these communal resources. Its homestead commission (*Höfekommission*) manages the *Hofrechte,* supervises the processes of sale and inheritance in the community, and establishes closed estates upon petition. It operates its own kindergarten and school, according to norms set by the German-speaking board of education of the Province. It communicates relevant decisions to the component households of the village, and it serves as a connecting link between the community and the Province of Bozen. Tret must depend for all such services and improvements upon the operations of the town council in Fondo. Whereas the communal council in St. Felix constitutes a formal body, representing the interests of the village to higher authorities, Tret must rely, in any attempt to win favors for itself, upon informal social contacts between the village and Fondo. Moreover, Trettners are often unable to reach a consensus. Thus, many decisions are imposed upon Tret by the town council, which can take actions that the *frazione* could not. These decisions are sometimes detrimental to the *frazione*, but need not be. For example, Tret benefited by the establishment of a dairy cooperative in Fondo, to which it sends representatives. Its own slovenly dairy (Italian: *caseificio*, Nónes: *kjasel*) closed its decrepit portals in 1961. Further, in 1969, it elected as representatives one Christian-Democrat and one Socialist, in order to maximize access both to the Christian-Democratic party in Fondo and to a promising socialist lawyer with political ambitions.

St. Felix also possesses a more elaborate ecclesiastical structure than Tret. The priest in St. Felix is a local resident, occupying a homestead (*Widum*). His role is very much that of *primus inter pares*, of a homesteader among other homesteaders, and as a result, he is deeply involved in the affairs of the community. Tret's priest lives in Fondo and divides his time between Tret, the deanery of Fondo, and other settlements within the commune. St. Felix, in addition, divides its population into four ecclesiastical organizations, the so-called *Standesbündnisse*, or status associations—consisting of married men, unmarried men, married women, and unmarried women. Each status association pays dues to the church, maintains a register of members, and hears special masses in the course of the year. Such associations are absent in Tret.

Thus, while a Felixer does not participate in an active social network of collaterals and affines, as does a Trettner, he is involved throughout his life in a series of formal organizations that socialize him to community and ethnic norms. A St. Felix child begins his association with such organizations when he enters the kindergarten, which was established in 1965 with the express purpose of strengthening Tyrolese identity. He spends eight years in school, under the supervision of teacher and priest, engaged in a curriculum that strongly emphasizes Tyrolese and German ideals. When he graduates from school, he moves into one of the ecclesiastical status associations that strengthen his belief in the values of the Holy Land Tyrol, and that encourage behavior appropriate to his age and station in life. As a young man he may join the fire-fighting squad, which involves him in a hierarchical structure of regional scope. Although St. Felix lacks the paramilitary organization of sharpshooters (*Schützenverein*) so characteristic of more affluent Tyrolese communities, and its band was disbanded in 1965, the fire-fighting squad has assumed many of the paramilitary aspects of these organizations. It requires wearing uniforms on festive occasions, soldierly bearing, military style in marching, and disciplined response to officers. It is notable that even in Tret enthusiasm for participation in the fire-fighting squad runs highest among the sons of German-speaking immigrants—but any pretense at military identification in the Romance village becomes the immediate target of communal humor. While Tret scoffs at local displays of militarism, St. Felix invests its fire-fighting squad with the aura of Tyrolese military traditions.

Finally, a man who lives up to local and Tyrolese expectations can count on election to the communal council—the formal repository of the community's autonomously organized life. In contrast, Trettners' support for organized participation in politics is always half-hearted and more than a little cynical.

This contrast originates not in the larger size of St. Felix, but in the different relationships between communities and the social and political structures beyond the village. A casual glance at another mountain village in Europe, Valdemora in the Sierra Maestra of Castile (Freeman 1970), shows that a community of only eighty-seven persons—much smaller than Tret—can support not only an autonomous communal council, but also two religious associations, corporate units of age mates, and feasting groups. The contrast between St. Felix and Tret appears to stem from their involvement with divergent cultural traditions. St. Felix replicates the Germanic pattern of autonomous organization of rural communities; Tret follows the Latin pattern that makes the rural settlements surrounding an urban nucleus dependent on the town. Here, St. Felix and Tret duplicate not only the wider patterns differentiating the South Tyrol from the Trentino (Dörrenhaus 1959: 76–80), but also the contrast between ancient Mediterranean and transalpine European urbanism (Ennen 1953).

CHAPTER XI

Cultural Confrontation

We have examined two communities in the Upper Anaunia: German-speaking St. Felix, and Romance-speaking Tret. In our attempt to compare them—ecologically, socially, and ideologically—we have noted numerous similarities and differences. In this chapter we wish to summarize our arguments, to evaluate what we have learned, and to ask further questions for which we do not yet have answers.

Ecological similarities between the two villages are notable. Both Tret and St. Felix are subject to much the same environmental imperatives and have met them with much the same technological response. Yet one is also struck by contrasts in the social and ideological realms. A significant difference lies in ideal patterns of inheritance. While homesteads in both St. Felix and Tret face essentially the same problem in assembling and maintaining a set of ecologically strategic resources, Tret delegates the decision-making power of many coheirs to one estate manager. St. Felix on the other hand, assigns management of a holding to the most available and willing son.

Ideal expectations diverge: notions of partibility here, impartibility there. Both, however, are not met in reality. There are indeed regions of the Tyrol where rules of impartibility are more or less realized, and where owners register their homesteads as "closed," that is, as impartible estates. There may also be regions where true partibility is in fact practiced. But in Tret and St. Felix, and—we suggest—in mountain peasant communities like them, rules of inheritance are in the nature of ideology and are not guidelines for action. As ideology, they have a referent other than the one to which they point ostensibly. We believe they refer to the concentration or distribution of legitimate power both within and among domestic groups. In St. Felix, an ideal pattern of impartibility legitimates the accession of one son to authority. In Tret, an ideal pattern of partibility legitimates a measure of continued influence on decision-making on the part of coheirs.

Stated in this way, we can see that ideal patterns of inheritance dovetail with the prevailing system of family organization, which is segmental in Tret, hierarchical and concentrated in St. Felix. The segmental family pattern in Tret, in turn, forms but one aspect of a widespread and wideranging network of generalized reciprocal relations. These relations even extend overseas, to the Americas, where relatives and friends continue to form part of an active network. In St. Felix, each homestead is pivoted upon a single heir, who has authority over all members of the household. Each household also monopolizes the activities and loyalties of its members, and hoards them in opposition to outsiders, even if these outsiders are related by ties of kinship.

The contrast between patterns of authority distribution entails still another dimension—the relation of authority patterns within the domestic group to the structure of the wider political field. Not long ago anthropologists tended to treat questions of kinship quite separately from questions of political organization, and to interpret kinship structure as divorced from the political matrix. More recently, however, there has been greater interest in tracing the effects of political ordering upon kinship organization. Thus, William Stephens (1963), in a cross-cultural study of the family, drew attention to the correlation of authoritarian types of family with authoritarian types of political systems. Yehudi Cohen (1969) has attempted to show in detail how different kinds of states seek to regulate systems of kin affiliation. Our data from Tret and St. Felix concur. The concentration or diffusion of authority on the level of the domestic group is affected by the nature of the political system and in turn influences it.

In this regard, Tret continually establishes new social linkages through the manipulation of its flexible network of kinship ties, but that network, in and of itself, bears no regular relation to the political field. Political pressures do not evoke a collective response, but travel through the social network, and are accepted, rejected, or circumvented by individuals in that network. Matters of marriage, inheritance, or managerial decision-making are all under the jurisdiction of the state, but it is individual behavior that matters. To the extent that the field of social relations is structured largely along lines of consanguinity and affinity, all social relations are *private*.

This is not so in St. Felix. From the very beginning, the St. Felix *pauer* has a dual role: he is patriarch within the domestic realm, and public representative of an organizational unit within the community. That is, he plays a certain role within the juro-political domain because of his private status in the social and economic sphere. In this sense, there are no completely private acts in St. Felix; all acts resonate in the public sphere and call into question the identity of the actor as a proper *Tiroler*. Thus, the structure of the St. Felix homestead articulates with the juro-political structure, whereas the social network of Tret does not. Just as the Nuer lineage, analyzed by Evans-Pritchard (1940), generates and facilitates the escalation of private conflicts into public concerns, so does the Tyrolese homestead. Conversely, it responds to influences from the political order outside the community. St. Felix is a politicized community; Tret is not.

We must seek an explanation of this contrast on three distinct, if interrelated, levels: on the level of Tyrolese experience, past and present, on the level of South Tyrolese experience within the Italian state, and on the level of relations between German speakers and Romance speakers within the Upper Anaunia.

There have been three major transformations in Tyrolese history. During the first transformation Bavarian invaders, led by a series of overlords, created the political unit of the Tyrol. Following Bavarian settlement and the absorption of previous settlers, the Counts of Tyrol were able to create a small, unified Tyrolese polity. In the construction of this polity, they granted and extended peasant freedoms, and thus secured peasant loyalty to their domain. While the mountain peasantry of Switzerland and of the French Alps forged a set of political alliances that extended horizontally from valley to valley, the Tyrolese won their freedom through membership in a hierarchically organized structure. In the Tyrol, homesteads came to form communities, communities formed estates, and estates formed the Tyrolese assembly.

In the course of the fifteenth and sixteenth centuries, loyalty to the Tyrol as an autonomous domain was transformed into loyalty for the Tyrol as an integral part of the growing Habsburg Empire. Although Tyrolese attempts to move toward political independence and religious freedom met with defeat, the Habsburgs allowed the Tyrol to maintain its own traditions within the imperial structure. The Tyrolese were to occupy a preeminent position within the empire, both as defenders of the true faith and as defenders of ethnic privileges. Loyalty to the Tyrol and to the Habsburgs implied an antagonism toward all challengers of Habsburg predominance.

In the course of the third transformation, the fate of the Tyrol came to be linked with the fates of other German-speaking groups within the Habsburg Empire. Just as the empire was riven by nationalist conflicts that linked groups within the empire to national units outside the imperial frontiers, so the Tyrolese began to experience a conflict between their role as bearers of an autonomous cultural and political tradition and their role as Germans. This conflict was to make its appearance in sharpest form with the rise of National Socialism in Germany, for it set Tyrolese who believed in the maintenance of their traditions "under God" against Tyrolese who believed in the leveling of traditions within the German realm, under the aegis of a unified polity.

In this conflict, many Tyrolese fell under the sway of the myth of Germany as a peasant nation. This myth, advanced by German nationalists since the beginnings of Romanticism, contained an implicit invitation to join in the construction of a new order that would not break with the past. The myth of *Blut* and *Boden*—the unit of German blood and soil—served not only to advance the notion of a *Reich* as an organic social and cultural unit in which all classes joined in a common purpose; it also told the peasants in the hills and forests that they were "somebody," that they had equal rights, even priorities, over city folk. The image of the peasant portrayed both a glorified past, and an auspicious future; it stressed not only what the peasant had been, but he would become. Thus, the German myth linked past to future, and served as a symbolic idiom capable of welding the various subgroups of German society into the new nation-state.

Over this period of a thousand years, therefore, Tyrolese experienced first the rise of their own autonomous polity, then inclusion of that polity in the Habsburg Empire, and finally fusion with an expanding German *Reich*. Each transition to a new level of political integration was accompanied by a parallel transformation of political loyalties. However, every movement toward incorporation in a larger system reinforced the continuity of the Tyrol as a political and ideological entity. The interests of Tyrolese as Tyrolese coincided sufficiently with the interests of the larger system for the Tyrol to keep its autonomy and identity.

While the rhythm of development held true for North and South Tyrolese alike, for the South Tyrolese its pressures and demands were heightened by forced inclusion in the Italian state. A population strongly rooted in a rural tradition was incorporated, in 1918, into a state created in the nineteenth century by urban elites, who had imposed their rule upon various rural populations, vowing to turn them into Italians. The South Tyrolese, with a strong tradition of self-government, found themselves subject to an alien, highly formalized government.

The Italian communes always focused upon an urban center; the Italian provinces and the Italian regions are not, generally speaking, based upon social or ecological realities. They are, in important ways, "empty" juro-political units—bureaucratic schemata superimposed upon a geographical grid. The state operates in the "Roman" mold, through formal bureaus set up by the urban center and entrusted with the administration of a law based on "theoretically constructed concepts of the person and his relation . . . denuded of any connection with concrete individuals and their acts . . . in their quest for 'imageless jural relations' [Wolf 1969: 291]." Alongside the formal bureaucracy there exists an army of middlemen, go-betweens, brokers—especially lawyers—whose task is to press the claims of particular parties against the general interests of the state. To the extent that the real, as opposed to the ideal, operation of the state depends upon the manipulation of its rules by this network of "friends of friends," the state is weak. It is the social networks that are real. Their effective operation weakens the rules: "the state of law becomes instead the state of the Machiavellians [Wolf 1969: 291]."

Within the Italian state, the South Tyrolese were exposed to a culture that has always regarded the city as the seat of civilization, and the countryside as subordinate to the civilized city. To the Italian, the rural dweller, the *contadino*, holds no honored place. Rural life is negatively valued as a way of life without profit and honor, a life one leads by necessity, not choice. Unlike the Germans, the Italians found it impossible to forsake a tradition of elitist art and literature in favor of a popular poetry, or to fuse the legacy of Roman and Medieval universalism with the new ethnic particularism. Moreover, far from glorifying the past, the Italians wished to break through a dismal legacy of decadence and foreign oppression. According to Croce (1933: 28), the carriers of the new Italian idea were not the *Volk* but the intellectual classes, "capable of thinking in historical terms."

The Italian Romantics did not create a popular literature, but a literature to educate the populace, "pedagogy in place of song welling upward from the profundity of the ethnic spirit, pedagogy in which the people were not teacher but student, not agent but patient [Croce 1933: 29]." True life is thought to be lived in the city, not the countryside. This distinction may be phrased in the terms Hannah Arendt has used in the discussion of freedom and servitude in Classical antiquity. According to her:

> . . . the prerequisite of freedom ruled out all ways of life chiefly devoted to keeping ones self alive—not only labor, which was the way of life of the slave, who was coerced by the necessity to stay alive and by the rule of his master, but also the working life of the free craftsman and the acquisitive life of the merchant. In short, it excluded everybody who involuntarily or voluntarily, for his whole life or temporarily, had lost the free disposition of his movements and activities [Arendt 1958: 12].

The three ways of life in which men might choose freedom all have in common their "full independence of the necessities of life and the relationships they originated [Arendt 1958: 12]"—the life of bodily pleasure, the pursuit of great deeds and words in the life of the polis, and the contemplation of things eternal. This, in the parlance of much of Italy, is not the life of peasants, but of *signori* (lords) and *preti* (priests), set apart in residence and manners from the life of the country people. But to the South Tyrolese, both noble and priest are next-door neighbors, attuned to rural life and its values.

The South Tyrolese confronted not only a state differing in organization, but one lacking in legitimacy, created—as they understood—by free thinkers and Freemasons in successful insurrection against the Habsburgs. They confronted a state that allowed them none of their own legitimating myths and customs, and that subjugated them to the rule of the godless city. To make the situation more grim, this state itself soon fell to a new set of rulers, the Fascists. The Fascist seizure of power had two effects on the South Tyrol. First, Fascism was a violent response to political weakness and disorder within Italy: through political centralization and the manipulation of political symbols, it sought to fill the "empty" bureaucratic structure with more tightly drawn linkages of legitimacy and loyalty, and, at the same time, rescue it from the day-to-day manipulation of network-oriented mediators. Second, the Fascist movement was strongly nationalist: it denied all ethnic minorities within Italy any linguistic and cultural identity of their own. In the case of the South Tyrol, this simply meant that efforts at forcible acculturation were now to be carried out by a more efficient administrative machine. The sharply increased migration of Italians into the province, as well as the introduction of heavy industry into the Bozen area, under the aegis of the Italian state, aggravated the trauma of separation from their traditional cultural and political matrix.

Under conditions of pressure and counterpressure, even simple acts—wearing the blue peasant apron to town—took on significance. The South Tyrolese insisted on the apron as a symbol of peasant status; the Italian authorities forbade it. Thus, the most trivial behavior was redefined through such antagonism as a defense of German folkhood. Adherence to custom and the preservation of German folkhood continue to be advocated through political and church meetings, through newspapers and radio. All mass media reinforce the symbolism of the terms "Tyrolese," "German," "peasant," and "Christian believer," defined by common opposition to a supposed Italian antithesis.

The parading of ethnic identity parallels, within the context of a modern European state, the process of "retribalization" so often observed in former colonial dependencies after gaining political independence. Abner Cohen has recently studied this phenomenon among Hausa traders in Nigeria. In the speeded-up "culture-building" and exacerbated ethnicity of this minority group he sees but,

> . . . the socio-cultural manifestation of the formation of new political groupings. It is the result, not of ethnic groupings disengaging themselves from one another, but of increasing interaction between them, within the contexts of new political situations. It is the outcome, not of conservatism, but of a dynamic socio-cultural change which is brought about by new cleavages and new alignments of power. It is a process by which a group from another ethnic category, whose members are involved in a struggle for power and privilege with the members of a group from another ethnic category, within the framework of a *formal* political system, manipulate some customs, values, myths, symbols, and ceremonials from their cultural tradition in order to articulate an *informal* political organization which is used as a weapon in that struggle [Cohen 1969: 2].*

* Originally published by the University of California Press; reprinted by permission of The Regents of the University of California.

Similarly, the South Tyrolese have responded to their position within the Italian state by emphasizing all the cultural contrasts that divide them from the Italian majority.

To the South Tyrolese, the rise of a Greater Germany to the north offered, for a time, the hope of a quick deliverance. During the 1930s the South Tyrolese were caught in a series of crosscutting pressures. Loyalties to a Habsburg polity, representing the ancestral political and religious identity of the Tyrol, soon found themselves at odds with the godlessness and ruthless abolition of parochial privileges characteristic of the National Socialists. Yet the Nazis stood for a maintenance of Germanhood, against the determined efforts of Fascist Italy to turn the South Tyrolese into Italians.

Within the South Tyrol, and within all local communities, these contradictions caused violent splits between those who opted for Germany and those who opted to stay at home—and, thus, within Italy. Rusinow has characterized the optants for Italy as "the urban middle class of Bolzano, those of the clergy who followed Canon Gamper rather than the Bishop of Bressanone [who opted for Germany], the landed aristocracy, and a minority of the peasantry [1969: 260]." Leidlmair notes among the resettlers a disproportionate representation of persons engaged in crafts and industry. Among the 75,000 who emigrated to Germany, the percentages of those engaged in crafts and industry was higher than those engaged in agriculture, and among the cultivators, owners of small holdings predominated over those with larger and impartible holdings (Leidlmair 1958: 76). But opting for Germany did not mean that either the optants approved of National Socialism or they indeed wanted to leave their homes: many optants simply hoped, in vain, that their declaration for Germany would cause Hitler to take a stand on their behalf against the Italians.

In a formal sense, the option signaled favor of Germanhood and the *Reich*. Substantively, however, the option represented a kind of rebellion. It was, first of all, a rebellion against those who wished to maintain a South Tyrolese identity within the framework of loyalties to a Habsburgian and Catholic past. It was, second, a rebellion of all sectors of the South Tyrolese population that were engaged in secondary and teritary pursuits, rather than agriculture, and that suffered most under the discriminatory policies of the Fascists.

Four factors have altered the situation of the South Tyrolese in the postwar period. First, the defeat of the Third Reich put an end to immediate hopes for a revision of political boundaries, and it motivated some form of accommodation with the new Italian republic. Second, the Italian state has abandoned, albeit grudgingly, the Fascist policy of forcible acculturation and granted much more cultural and political autonomy to the South Tyrolese. This has reduced the appeal of an *irredenta* for the South Tyrolese population. A terrorist movement aimed at gaining immediate independence from Italy, arose in the 1960s; while many sympathized with the terrorists, few were willing to actively support the movement, which was short lived. Third, about one-fourth of the 75,000 optants who actually departed for Germany during World War II have returned to the South Tyrol (Leidlmair 1958: 83), but the permanent loss of the other three-fourths has left the South Tyrolese population more rural than before, a trend further enhanced by the high birth rate of the South Tyrolese peasantry. This ruralization has reduced the number of persons who look to Germany for opportunities of employment and social advancement. Fourth, economic developments in Italy and growing prosperity in Europe have increased job opportunities within the Province of Bozen, thus offering a measure of economic stabilization in the Tyrol.

At the same time, however, there has been no easing of the insistence of South Tyrolese upon cultural and political separateness; if anything, this insistence has grown stronger in the context of a more permissive political atmosphere. The postwar period witnessed the rise to prominence of the South Tyrolese People's Party (*Südtiroler Volkspartei,* or SVP). Organized in opposition to both National Socialists and Facists during the last years of World War II, it emerged in the immediate postwar years as the sole political body representative of South Tyrolese claims within the province, the region, and the Italian state. Its politics have been generally "accommodationist," but its rhetoric is strongly nationalist. It no longer relies on appeals to Habsburg nostalgia, but sharply emphasizes the cultural and political distinctiveness of the South Tyrolese. The SVP insists on the need to maintain a united front against a common enemy. As a result, it has successfully discouraged the development of rival political groups within the South Tyrolese population; it has eased relations between Tyrolese who, during the war, opted for resettlement in Germany and those who stayed in Italy; and it holds appeal for the generation that lived its youth under the impact of the National Socialist ideology. Many of this war generation served in the German army during World War II in their early manhood. It is this generation, with its German nostalgia, that has replaced the older Habsburg-trained spokesmen, and now dominates most South Tyrolese communities. The success of the SVP, however, depends upon a continued favorable political context. An economic depression, for instance, or the renewed rise of fascism within Italy, would seriously impair the continued appeal of the SVP. While the South Tyrolese continue to accept a role within the Italian political system, they defend themselves against social and cultural encroachment through reliance on a traditional set of cultural responses.

We are now in a position to return to the geographical focus of our study, and see how these processes have affected the Upper Anaunia and—specifically—the villages of St. Felix and Tret. Ecologically similar, the two communities diverge in family structure and in the ways domestic units fit into the wider social structure. This divergence, we believe, stems from the villages' differential involvement over time with the political systems that have exercised sovereignty in this marginal mountain area.

It must not be supposed that St. Felix and Tret confront each other as opposing camps of German folkhood and *Italianità* in every social, economic, or religious context. In fact, throughout most of the year, interaction between them is matter-of-fact and, on an interpersonal level, often pleasant. Ethnic affiliation is relegated to the background when Trettners and Felixers visit each others' churches, play games of cards at the inn, close a deal over a glass of wine, or chat in the marketplace in Fondo. Multilingualism makes easy communication over a wide range of common interests. Yet, all these interchanges take place in the public eye. Backstage, to use a metaphor of Erving Goffman, within each community villagers communicate among themselves, in terms of a set of fixed images of the other population. These stereotypes may not apply to any one member of the population so characterized, but they do reflect historical experiences each group has had with the other. In other words, stereotypes are the folk equivalent of the anthropologist's structural and historical approach: they differ from it by seeking relevant explanations in peoples' "nature." To the extent that they do not question the nature of the boundary, stereotypes are masks, myths. Yet, they do voice present and past concerns.

We have spoken of the emphasis that Felixers place on *Ordnung,* order and orderliness. They regard Trettners as disorderly, dirty, and lazy. Such an opinion is quite unwarranted. We can certainly point to neat and orderly or slovenly and disorganized households in either village. The imputation of dirtiness clearly has social–psychological functions in projecting negative traits upon the opposing group; it should also be read as a reflection on the abilities of women drawn from the other group and, hence, on their desirability as marriage partners. Similarly, the Trettners work quite enough; but it is true that less of their labor goes into caring for field and livestock, as less of their life is closely bound up with agriculture. The Felixers further characterize the easy sociability of the Trettners as "talking too much," and their ready expenditure for modern articles as "spendthrift."

Trettners reciprocate by calling the Germans "stingy," accuse them of "living in caves"—a comment on their continued use of traditional housing—and say ironically: "They work too much, with too little results. If the border between Italy and Austria is finally drawn between Tret and St. Felix, they'll starve to death and we will make money by selling them food." The Germans emphasize the Trettners' lack of military prowess; the Trettners recognize the effective militarism of their neighbors, but say derisively: "If you tell a German to go through the wall head first, he'll go through the wall head first."

The different valuations of orderliness, discipline, and soldierly bearing came through quite clearly when informants from the two villages looked at photographs taken at their annual processions around the village boundaries. "Look at those Italians," said a Felixer, looking at the picture taken in Tret, "walking along just like a herd of sheep." "Look at those Germans," said a Trettner, on noting that in the picture of the procession coming down a rocky and uneven mountain path, all the participants were in step: "They're all in step. Isn't it just like them!"

Neither Trettner nor Felixer trusts the other. The Germans realize that trust for the Trettners is a relative matter outside their social networks. The Trettners have negative memories of the sudden outbursts of cruelty and nationalist ferocity on the part of their neighbors. During the war, many Felixers denounced Trettners hiding in the village to escape from forced labor service or military obligations to the German field police. "We have always known that they are all S.S. men," commented one villager, upon hearing of Adolf Eichmann's capture by the Israelis. While Felixers take credit for capturing escaped Allied airmen during World War II and handing them over to the German police, Trettners feel strongly that a man, once escaped from bondage, should not be handed over to organs of the state under any circumstance. Some men in Tret, having suffered forced labor or other indignities during the war, will have nothing to do with Germans and refuse to drink German wine; people in St. Felix retaliate by not patronizing the store of a man reputed to have been the local partisan leader in World War II.

"Different valuations of orderliness, discipline and soldierly bearing": local firemen in uniform march in processions in St. Felix (above) and Tret (below).

In this cycle of overt and covert insults, the Felixers are apt to give voice to their sense of superiority over their *Welsch* neighbors, a stance that derives from their historic participation in an expansionist economic and political field. The Trettners are often ironically aware of the differential in specific gravity, economic and political, between the German and the Italian orbit; at the same time, they take some malicious pleasure in the discomfiture of their German neighbors, who must work within the framework of an Italian state. Thus, while individual encounters readily pass over the ethnic boundary, the ready and ever-present stereotype continuously re-establishes it.

Trettners rely on a social network that on occasion crosscuts the political domain; but it is the network of social relations that guarantees security and advantage, not the political order. The state exists "above" and "beyond" the social network. It receives little loyalty and is granted relatively little legitimacy. This was true in the period of the Habsburg Empire when the Trentine peasantry showed little interest in the Italian *irredenta,* and it is true today in their attitude toward the Italian state. In contradistinction to their German neighbors, the *todesc,* Trettners will indeed identify themselves as Italians, but Italianhood is a stance assumed only in contrast to Germanhood; it has little substantive reality of its own.

For the Nónes, Italians are still "those other people" whose historical associations are with the old Kingdom of Italy, and in common parlance, Italy is still more often the area of the *Regno* than the state that exercises sovereignty over the territory of the Italian republic. An immediate cause of this attitude lies in the nature of the Italian state itself, which activates its formalized bureaucratic framework through the activities of political actors organized in informal networks. Realistic influence in this state thus consists in finding the right person to exercise his influence on your behalf, as Trettners did in 1969, when they granted their votes to candidates of two different parties in order to have leverage in both camps. This realism in seeking personalized connections does not go hand in hand with any strong commitment to the operations and maintenance of the state. Tret conforms to the picture of Italian political culture depicted by Almond and Verba as "one of unrelieved political alienation [1965: 308]." One relies on people: the state is seen as an antagonistic force.

Yet the immediate cause of the relation between the Nónes and the Italian state must be sought not only in the present, but also in the past. We have noted that Nónes history was characterized, during the centuries following the first millenium, by some attempts to generate intercommunal federations, attempts that were forever upset by shifting factions, coalitions of factions, and factional fights. Unlike similar populations in the Swiss or French Alps, however, the Nónes never were able to create a unit within their valley strong enough in resources to achieve any sort of enduring dominance. At the same time, interference from outside, by Bishop and Count, continuously fed factional disputes and inhibited further consolidation. As a result, consolidation came only after 1450 through the imposition of outside rule, first by the Counts of Tyrol, later by the Habsburgs, and not through the development of a political structure generated within the valley.

The inability to provide a modicum of internal political organization also inhibited the development of an independent Nónes cultural identity. Moreover, the imposition of a political framework from outside also served to draw into the service of the Habsburgs many a local notable who could have contributed to the formation of a Nónes political and cultural entity. Their isolation by German and Italian speakers from kindred linguistic groups in Switzerland, Gröden, and Friuli still further cut them off from efforts to maintain linguistic integrity and to develop the literary creativity characteristic of those areas. Thus, the Nónes could maintain, until the twentieth century, an intimate sense of themselves as people "like us"—who speak Nónes at home—but who have always lived within a political system not of their own making. Today, even the sense of linguistic separateness is under challenge, as the present generation begins to transact most of its external business in Trentine or standard Italian. There are still a few people in Tret who keep books of poetry written in Nónes in their houses, but most villagers claim to be unable to read Nónes verse when it occasionally appears in local newspapers or calendars. Some individuals, in fact, have given up the use of Nónes entirely.

To the extent that Nónes continues to be the language of family life and, hence, provides a basis for a sense of common identity, the Nónes bring to mind another mountain population within the Mediterranean area, the Berbers. We are thinking here not of the politically autonomous Berbers of the "zone of dissidence," who are able to withdraw altogether from the power of the state into their own zones of refuge (Gellner 1969). Nor are the Nónes like those Berbers who accepted a specialized military role within the structure of the *makhzen,* the power center of the state (Schorger 1969). Rather, the Nónes call to mind those Berbers who, unable to escape the power of the state and not equipped with any special privileges, survive interstitially on the subpolitical level within modern Moroccan society (Vinogradov 1970). When the state is weak, they play at factional politics for fragments of lost power; when the state is stronger, they exercise an ability for playing patron–client for local privileges and exemptions. They maintain their flexibility in action by committing themselves neither to the structured oppositions of a segmentary sociopolitical system, nor to enduring relations with the power holders of the settled zone. Through their linguistic identity as Berbers, they can defend an inner circle of intimacy against outsiders. At the same time, they can identify themselves as Arabs when the occasion demands.

Similarly, the Nónes maintain their flexibility in action by avoiding commitments with one or the other side, and they are able to trim their sails to the changes in the prevailing winds. We can think of the Nónes as a kind of "submerged Berbers" and surmise that there are indeed many such populations in Europe, who are never able to fully insulate themselves from the state but who, nevertheless, present a barrier of ethnic identifications to state interference.

In their flexibility and nonalignment with the state, the Nónes contrast sharply with their German-speaking neighbors. We have seen in St. Felix that the structure of the homestead lends itself to a process of political escalation, and we are now in a better position to understand the purposes served by that escalation. We have found that the two populations resemble each other markedly in their ecological adaptation to the local environment, but we have noted how they differ in the ways in which authority is wielded within the two communities. We have also begun to understand some of the ways in which this authority is integrated into the operations of the wider political system. We have related the nature of the Nónes social network to the character of their historical experience; we can do the same with the German speakers of the *Deutschgegend*, a marginal area within the South Tyrol.

We must not forget that the German speakers entered the area south of the Brenner as colonists and advanced against zones previously settled by Romance speakers. This competition for living space led to the elaboration of their political and military linkages, thereby providing an organizational infrastructure for the wider polity of the Tyrol. This infrastructure also served them well in their advance into frontier areas. The expansion of the German speakers into the Anaunia resembles that of the pastoral Nuer, pressing upon their pastoral predecessors in Nuerland, the Dinka (Sahlins 1961). The

Dinka were the first settlers in the area: their social organization was of shallow lineages, with no need of any wider political or military linkages. The Nuer, on the other hand, had to marshal all their forces against the Dinka. On the basis of a system of social organization originally very much like that of the Dinka, they built up a system of wide-ranging lineages, which could mass large numbers of men against the Dinka. Just as the Nuer possessed a means of expanding organization from the level of the camp to that of the local descent group, and beyond to the level of the political lineage, so did the Tyrolese develop a framework for organizational escalation reaching from the homestead through the commune and the peasant estate to the Tyrolese polity. Manned by a free peasantry, with the right and obligation to bear arms, this organization was at once social, political, and military. It provided a viable apparatus for local self-management; it acted as a shield against previous Romance settlers; and it facilitated mobilization against potential opponents from the outside.

Such an infrastructure served them well not only in confrontations with peoples settled in the great longitudinal valleys, but also in their advance into unsettled transversal valleys and upland frontier zones, when they were to come into conflict with earlier inhabitants. Such an upland frontier area was the Upper Anaunia. St. Felix and Tret were both founded as advance posts for two movements of expansion—one advancing southward from the Meran and Lana over the Gampen Pass, the other, northward from Fondo. Meeting in competition over plowlands, pasture ground, water, meadows and forest land, each sought wider ties of alliance in the areas to the south and north. But while Felixers could link up with a viable political infrastructure already established in the adjacent valleys and piedmont, the Nónes in general and Trettners in particular found only weak support in a fragmented and impotent political structure, overshadowed by the more dynamic polity to the north. Thus, the Trettners put their reliance in ties of affinity, weaving and reweaving multiple interpersonal linkages. The Felixers, however, as an advance guard into contested ground, could rely on the Tyrolese apparatus developed first by the Counts of Tyrol and later by the Habsburgs. Although ecological conditions in the high mountains allowed them to duplicate the cultural forms of the more prosperous lowland and piedmont only approximately, they could direct their local social organization into a more complex and viable political framework. In their local culture, ecological adaptation and commitment to Tyrolese statecraft fused with each other. In turn, participation in statecraft reinforced their cultural antagonisms to other ethnic groups. Thus, St. Felix, like other Tyrolese communities, experienced the vicissitudes of the peasant wars, the wars against the French, the efforts to contain the Italian *irredenta*, and the trials of World War I, not only in its own terms, but also as a participant in a common Tyrolese fate.

Like other South Tyrolese communities, St. Felix suffered the fate of being severed from the homeland and transferred to Italy. In the case of St. Felix this was doubly bitter, since in the period between World Wars I and II it was assigned to Fondo and, hence, to the Italian-speaking Trentino, rather than to the province of Alto Adige, with a German-speaking majority. Yet in spite of these pressures, the village effectively retained its German identity, even under the onslaught of Fascism. Like other South Tyrolese communities, Felixers had to face the option, and as elsewhere the choice between moving to Germany or staying in Italy bitterly divided the inhabitants. Like marginal mountain peasants elsewhere, most chose to move to Germany. Reflecting on the option some twenty years later, one Trettner said: "Of course they voted for Germany. They saw a chance to sell their homesteads and get out. Nobody before or after has ever offered them money for those miserable holdings and caves in which they live." Those who voted for Germany saw in the *Reich* opportunities for a more viable existence and for social mobility among people with whom they shared a language and culture. Few actually left, but throughout the war, and especially during the German occupation of the South Tyrol, St. Felix was pro-German.

When the dream of the Third Reich came to nought, Felixers again had to adapt to living with their Italian neighbors in an Italian state. Since World War II, they have been strongly loyal to the South Tyrolese People's Party; they vote solidly SVP and follow its political and ideological promptings. We have already discussed some of the reasons for the success of the SVP above. In St. Felix, as elsewhere, Habsburg-oriented loyalties have declined with the demise of the Habsburg Empire and the decay of the Habsburg-educated generation. Reunification with Austria, now represented by the small Austrian republic, seems increasingly unlikely.

The generation now politically in charge of St. Felix is the generation that was recruited into the German army during World War II. While its outlook is strongly colored by participation in the war, much of that experience is now irrelevant. At the same time, the expanding economy of province and region offers a greater opportunity than ever before for a larger percentage of Felixers to stay at home as members of the village core population. Politically, control of the communal council has moved from the older political actors to younger men, recruited from full-time and part-time peasant owners and heirs alike. The village therefore stands in need of modified collective representations. The Tyrolese myth, which has successfully assimilated and adapted to three major transformations in the past, requires revision. To date, the South Tyrolese People's Party has been successful in providing these revised collective representations. It accepts accommodation with the Italian state, but only if that state recognizes the political autonomy of the South Tyrol. It celebrates the conditions of peasant existence, thus supporting the efforts of the Felixers to retain roots on the land. Moreover, it underlines at every opportunity that the South Tyrolese as Tyrolese are true Germans. These ideological messages are transmitted to St. Felix by the communal council, which is solidly SVP: local wits refer to the councilmen as the G'scheiten, the clever ones. The new kindergarten in St. Felix, erected with outside funds collected in the German-speaking countries to the north of the Brenner, is a living monument to this cultural stance. The effect of internal political mobilization and of external support has been, once again, to emphasize the structural barriers dividing St. Felix from communities like Tret.

St. Felix is not only a more politicized community than it was in the past; its young families are also more committed than before to leading a life on the land. They are making more effective use of the land, and extending the scale of their operations into the Trentino. We have seen that in their efforts to mechanize agricultural equipment, they are postponing other expenditures that would link them more firmly to the consumer economy. In this regard, their sense of identity as members of a free and active peasantry is reinforced by the political messages of the SVP. Their specialized, adaptive strategy to a local ecological context meshes neatly with a political appeal concerning the structure and organization of the wider polity. That political appeal underlines the identity of the South Tyrolese as rural, patriarchial, Catholic, engaged in maintaining age-old patterns of local self-government. Felixers thus fortify their grip on the land, their resistance to urban linkages, their maintenance of authority patterns within the community, and their autonomy from the mass-oriented party politics of the Italian political system. The symbols used by the SVP evoke a separatist tradition. The Felixers manipulate these symbols to grant legitimacy to the narrow panoply of ecological tactics that they employ in the immediate context of the village.

The Trettners, in turn, are rapidly becoming urbanized. For them, the ideal life is not on the land, but in the city. Virtue does not lie in wearing the blue peasant apron, the *Schurz* of the Germans or the *grembiale* of the Italians (Nónes: *grembiúl*), but in a style of life that suggests the possibility, if not the reality, of seignorial leisure. In Hannah Arendt's terms, they want to approximate the freedom that lies in the life of the polis, and the ideal location of that life in the twentieth century is not Italy, but the United States. Migrants to the United States are free; those who return have also returned to a life in fetters. Yet even the unfree have freedom to copy the ways of the free—the freedom of fashion. Trettners are therefore interested in fashion and fashion change. The messages that reach Tret from the outside world, in the letters and gifts sent by relatives, and from magazine photographs, serve to stimulate the Trettner to remake himself in another image, in hair styles, clothing, household goods, musical tastes. The villagers participate in the outside world through the mass media and commodity circulation, purchasing such goods as are obtainable in regional stores at prices they can pay. Consumer goods are points in a ritual game. The game consists in an external approximation to the expressive behavior of pacesetting cosmopolitan elites, an attempt to assert a social and economic equality denied by the reality of peasant life. Thus, St. Felix and Tret differ not only in internal structure and in their external relations to larger polities. They are also engaged in essentially distinct symbolic games.

We see in this analysis how two social and cultural microcosms have developed divergent structures, while yet engaged in close spatial and social relations. Our study thus supports the statements made recently by Fredrik Barth in an introduction to a book on ethnic groups and boundaries, where he emphasizes that boundaries are created and persist despite a flow of personnel and social relations across them (1969: 10–11). We are sympathetic to his view that since it is the ethnic boundary and the conceptualizations of a people themselves about the boundary that define interaction between two ethnic groups, one must direct attention to understanding the ethnic boundary. We have followed an essentially similar approach in our study. Yet we have found that the actions of people at the local level—particularly with regard to interlocal contact across the ethnic boundary—do not respond only to local influences, but are affected by actions and ideals of a much wider area. The competition between Felixers and Trettners over available resources and the means of using them has resulted in a structural heightening of differences between them, much as if both had responded to Lévi-Strauss' principle that two players in a game whose resources threaten to become uniform deliberately introduce differences in each of their games (1952: 46). More important, however, the villagers seem to have followed the second tactic for structural differentiation defined by Lévi-Strauss, which is "to bring into the game, whether they will or no, new partners from outside, whose stakes are very different from those of the parties to the original coalition [1952: 47]." The Felixers deliberately broadened and widened their connection with the German culture sphere to the north, the Trettners with the Italian society to the south.

Here we must part ways with Barth. He argues rightly that "people's categories are for acting, and are significantly affected by interaction rather than contemplation [1969: 29]." However, Barth uses this point to deny the utility of an approach in anthropology that

> . . . has been to dichotomize the ethnographic material in terms of ideal versus actual or conceptual versus empirical, and then concentrate on the consistencies (the 'structure') of the ideal, conceptual part of the data, employing some vague notion of norms and individual deviance to account for the actual, statistical pattern [1969: 29].

Our account has repeatedly emphasized statistically analyzable patterns; but it must also, in the nature of the material, come to grips with the facts that the two populations themselves "dichotomize the ethnographic material in terms of ideal versus actual," and that their involvement with different norms has important consequences in their daily lives. These norms specify ideal access routes to different systems of social and cultural interaction. Felixers need not have worked in Germany to feel German, but they in fact identify with Germans and will not work in Italian industry. The significant channels of communication—communication of ideas about the shape of existence now and in the future—orient them toward quite different economic, political, and religious systems (see Deutsch 1966).

In principle, the point is similar to the one made by Marshall Sahlins (1965a) on the disjunction of descent ideology and local group composition in primitive societies. The ideology of descent responds to the functional context of *political* alliance and cleavage; one cannot argue from an ideology of descent to the facts of recruitment into local groups. What Sahlins says of descent structure can be applied at the same time to nationalist ideology:

> The overlying descent structure is no expression of the underlying descent composition. Something to the opposite: the major descent system orders genealogical facts in allegiance to its own principles. Subject to arbitrary interpretations, ancestral facts appear subordinate to doctrines of organization, not the doctrines to the facts. Wildly different descent groupings are contrived from fundamentally similar descent aggregates. Truth in a descent system is a specification of structure, not of birth, and in this event 'true ancestry' may assume the status of a sociological lie.
>
> . . . In the conventional wisdom, structure is the precipitate of practice.
>
> . . . Descent is not recruitment, but arrangement and alignment, in the first place a principle of political design, exercising arbitrary constraints on the suppositions of ancestry.
>
> . . . modes of recruitment and affiliation at the lowest levels, these may merely reflect higher-level structural imperatives, thus injecting into everyday household and community life the relations appropriate to the working of the larger polity [1965a: 106].*

As in descent ideology, so is this the case in the ideology of nations. Germany and Italy as nations are specifications of structure, "not of birth," and indeed affiliation with one or the other may have the status of a sociological lie, perpetrated in the interests of political design. Moreover, ideological statements of affiliation with one or the other structure, whatever the nature of their sociological "truth," have real effects at the level of household and community. This does not mean that Tret or St. Felix displays all the attributes— in microcosm—of Germany or Italy. There is, however, communication between macrocosm and microcosm, in terms of ideas and messages about ideas, and such communication is about real things in a real world: differential access to life chances, differential modes of acting on the public stage, differential power.

The same point may also be applied in the study of ethnic groups. Ethnicity, as Abner Cohen has shown in his study of the Hausa in Nigeria (1969), is politics. The Nónes are losing their identity even as a linguistic community, because they have not generated a social structure geared to competitive mobilization within a larger political field. The Tyrolese north of the Brenner, in Austria, are rapidly abandoning traditional cultural patterns in favor of orientations that will serve them more effectively to enter an expanding industrial and commercial German and Austrian world. Traditional folkways are resurrected only in a self-conscious and staged manner,

* Reprinted by permission of the Royal Anthropological Institute of Great Britain and Ireland.

in the interests of the tourist trade. The South Tyrolese, however, rely on ethnic politics to defend their interests within the Italian state against more powerful opponents; hence, they elaborate the symbolic markers and content of a South Tyrolese culture to offset their relative political impotence. St. Felix, like other South Tyrolese communities, is caught up in this movement of cultural elaboration and cultural defense. It responds eagerly to an ideological appeal that has only tenuous connection with its ecologically grounded system of action. Yet, in its own way, its ideological commitment is quite real. It is real, most certainly, "in its consequences."

We began our study in the tradition of cultural ecology, attempting to relate the life situations of the people of St. Felix and Tret to the environment. We discovered that this approach yielded a measure of understanding of the similarities between the two villages, but we soon found that to understand the ways the villages differ we needed to examine their relations to a larger social and cultural environment. In recent years, ecological anthropologists have increasingly stressed the importance of the intergroup environment as a determinant of local situations, but we found the insights offered by ecological anthropology inadequate to explicate intergroup relations in the Upper Anaunia—a region characterized by great political and ideological history.

This experience has raised doubts in our minds about the usefulness of ecological anthropology in the study of complex societies. First, it has been most successful in the study of relatively isolated, primitive societies. These can be treated as more or less static, self-regulating systems, as can attempts to delineate the mechanisms of negative feedback through which homeostasis is maintained. Most complex societies, however, are not static, but dynamic. They are not characterized by a dominance of mechanisms of "negative feedback," but rather by processes that utilize "positive feedback," often resulting in "oscillation," or disorder in the system. They are not, in Lévi-Strauss' terms "cool" systems, but "hot." Therefore, such studies do not tell us much about communities like St. Felix and Tret.

Second, because the study of homeostatic systems has been emphasized in ecological anthropology, the only kind of change that this discipline has considered is systemic divergence through movement into different microenvironments. In such studies, one can see how adaptation to a new microenvironment results in a new homeostasis. This kind of analysis is most applicable to the process of fission in isolated primitive societies; it is least applicable to the process of fusion and synthesis that governs the rise of complex societies. Complex societies are ecologically grounded, but the rise of the state introduces into the ecological set a specifically political element that transforms problems of ecological limitations into decisions of a *political* economy. State development not only entails the subjugation and incorporation of various locally adapted groups. It also aims at the development of an apparatus that recombines ecological resources in new and unforeseen ways. The state strives to create, above all, in Eisenstadt's words,

> . . . free-floating resources, i.e., resources—manpower, economic resources, political support, and cultural identifications—not embedded within or committed beforehand to any primary ascriptive-particularistic groups. It also created a reservoir of generalized power, in the society, not embedded in such groups, that could be used by different groups for varying goals. These free-floating resources and the generalized power were both necessary . . . for the rulers to establish autonomous political institutions and to pursue some autonomous and differentiated political goals and activities; and they created the potential for the institutionalization of differentiated political roles and organizations" [1963: 27–28].

We, therefore, cannot understand the microecology of St. Felix and Tret without raising questions about the changing economic and political systems which encompass them and to which they must respond. The institutions of market and state relate to local ecological processes, but they are not explicable in microecological terms. The market may indeed be seen as a set of mechanisms designed to challenge local ecological adaptations and reallocate resources in ways unique at the local level. There is a specific sense in which the market is antiecological, as Karl Polanyi has pointed out:

> As the development of the factory system had been organized as part of the process of buying and selling, therefore labor, land, and money had to be transformed into commodities in order to keep production going. They could, of course, not be really transformed into commodities, as actually they were not produced for sale on the market. But the fiction of their being so produced became the organizing principle of society. Of the three, one stands out: labor is the technical term used for human beings, in so far as they are not employers but employed; it follows therefore that henceforth the organization of labor would change concurrently with the organization of the market system. But as the organization of labor is only another word for the forms of life of the common people, this means that the development of the market system would be accompanied by a change in the organization of society itself. All along the line, human society had become an accessory of the economic system [1957: 75].

But, while production could theoretically be organized in this way, the
commodity fiction disregarded the fact that leaving the fate of the soil
and people to the market would be tantamount to annihilating them
[1957: 131].*

As we have already seen, the state is a machine of "free-floating
resources . . . not embedded within or committed beforehand to
any primary ascriptive-particularistic groups [Eisenstadt 1963: 27]."
The industrial and commercial states of the present day aim at the
creation of systems in which the factors of production, including
labor, are rendered mobile and "free floating" so that they can be
allocated and reallocated to any part of the system, and in which
patterns of consumption become relatively standardized. Politically,
the modern state hopes to create populations whose loyalties and
commitments are no longer primarily to locality and region, but
to the system as a whole.

The agents of national economic and political mobilizations are
elites that win favor for their activities by combining claims to
economic resources with appeals for political support. Appeals for
political support, in turn, require development of the appropriate
symbols and codes of behavior that can elicit a positive response
from potential partisans. Contending elites therefore not only compete for resources and power, but also vie with one another in the
elaboration of symbolic systems that can "win the hearts and minds
of the people." These symbols and codes of behavior, moreover,
must form part of an ideology that defines the goals to which economic allocations and political support are to be put. Competition
between elites—for resources, for allies and followers, and for
symbols—creates a political ecological system of relationships that
depends for its very existence on its ability to manipulate microecological adaptations. If we are to understand what happens in
villages within complex systems, therefore, we need not only a
better understanding of political economy, of the processes of
economic funding of power capabilities, but also one of political
ecology, of the system of relationships between groups possessed
of differential access to resources, power, and symbols.

The competitive interplay between elites and with groups of
potential supporters generates the cultural system we call the nation.
The nation, in turn, serves as a framework within which such interplay can go on. Nations are not ready-made. The genesis and
development of a nation has a history: it happens under specific

* From *The Great Transformation* by Karl Polanyi. Copyright 1944 by Karl
Polanyi. Copyright © 1972 by Marie Polanyi. Reprinted by permission of Holt,
Rinehart and Winston, Inc.

circumstances, encounters particular obstacles, and develops characteristic responses to these circumstances and obstacles. The historical process bears upon different groups and different regions at different times and with differential intensity. The synchronization or failure to synchronize developments separated in time and space has important consequences for the next phase in the movement of the system. Moreover, the rise of a nation and the characteristic rhythm of its rise depends both on internal factors and on the external relations obtaining between national systems as a whole, or between elite groups belonging to different national systems (Wolf 1953).

These processes operate at one and the same time on the level of the nation and on the level of constituent subsystems. It is on the level of local microsystems that anthropologists can most easily study and analyze developmental sequences, as we have striven to do in the study of St. Felix and Tret. To the extent that local systems participate in a wider and more significant drama, we have also questioned the factors that determine incorporation of the subsystem into the larger whole. Our account has delineated some of the variables that are significant in this process: the position of a local subsystem in the process of ecological succession; the degree of development of a local and autonomous political structure; the timing of the emergence of such a structure in relation to more wide-ranging and inclusive political units; the degree of "fitness" in linkage between the local structure and the structure of the wider political field; the ability to transform symbols pertaining to the local structure into symbols pertaining to the wider structure. In this regard, St. Felix and Tret serve as examples of two types of nation formation. St. Felix is typical of those cases in which an existing political structure is transformed to fit the larger structure of the nation; the Trettners, in contrast, were "nationalized" by a formal procedure, with their own networks overlaid by a functioning, bureaucratic national structure.

Yet our study also teaches us that these differences are not "inherent" in the culture of the two populations. Each village responded differently at different times, to the interests and demands of particular elites within a wider political field. This field takes on its characteristic form in response to the activities of various elites; its shape and content will vary with the rise to dominance of new elites, or with the growth of coalitions among elites (Moore 1966). The nation-building process "on the ground" responds to political developments. Its impact on the Anaunia differed under the Habsburgs, the National Socialists, the *irredenta*, the Fascists, and the coalition ruling postwar Italy. Contending elites will produce variant forms of the invitation that they extend to the population at large; variant forms will mobilize different groups of followers. The invitations to nationhood sponsored by groups led by a Bismarck or Cavour will have different structural implications for nation-making at the village level than those sponsored by a Lassalle or Garibaldi. What happens at the local level then determines what options are open to the larger polity in the future.

Population Statistics

The investigator who wishes to make use of official and unofficial population statistics on the South Tyrol and the Trentino soon finds himself embroiled in all the artifices which statistics can present to the unwary. This is compounded in the present case by their common use in various nationalist arguments. Among the difficulties in using statistics drawn from various national censuses are the following:

1. The Austrian census of 1910 classified populations by "language habitually spoken" (*Umgangssprache*), not by mother tongue, giving rise to the possibility that people with German as their mother tongue could be classified as Italians, and people with Italian as mother tongue could be classified German speakers.

2. The Austrian census of 1910 included in its statistics government and military personnel on temporary duty in the area, but excluded migratory Italian laborers present in the South Tyrol at times other than during the census.

3. The Austrian census did not distinguish between Ladinsh speakers and Italian speakers.

4. Italian censuses from the Fascist regime produced highly unreliable nationality statistics; for example German-speaking South Tyrolese with Italian family names were frequently counted as Italians.

5. Italian censuses did not distinguish Ladinsh speakers from Italian speakers. On the other hand, when—in 1939—Ladinsh speakers were allowed to opt for Germany, along with German speakers, Italian agencies classified Ladinsh and Germans together as *allogeni*.

6. Many discussions of population movements in the area do not take account of changing assignments to one or another province of the following communities: Unser Frau, St. Felix, Proveis, and Laurein in the Anaunia; Truden/Trodena and Altrei/Altidena; and the district of Ampezzo.

Fiebiger (1959: 15–16) has systematically corrected for these territorial changes. We reproduce his population figures here, changing only the name employed by him, "Welschtirol," to "Trentino":

POPULATION FIGURES FOR THE TYROL AND THE TRENTINO

Year (A.D.)	North Tyrol		South Tyrol		Trentino		Total	
	Number	(%)	Number	(%)	Number	(%)	Number	(%)
1754	189,000	(31.9)	198,000	(33.4)	206,000	(34.7)	539,000	(100)
1835	209,000	(29.0)	220,000	(30.7)	290,000	(40.3)		
							719,000	(100)
1910	275,000	(29.0)	285,000	(30.1)	387,000	(40.9)		
							947,000	(100)
1951	389,000	(33.3)	383,000	(32.7)	398,000	(34.0)	1,170,000	(100)

For a discussion of population problems before 1754 we refer the interested reader to Wopfner's excellent presentation and discussion of available data (1954, Vol. II: 222–324). We have adopted Wopfner's estimate of a population of circa 500,000 for 1650. We also accept Wopfner's rough estimate of the rates of increase in the number of *homesteads* 1312–1427 and 1427–1837. He suggests that homesteads increased by one-half, between 1312 and 1427, and doubled again between 1427 and 1837 (1954, Vol. II: 223). These estimated rates are based on tax lists collected previously by Otto Stolz from a number of communities for the dates indicated. The tax lists do not indicate the criterion for the inclusion of the homesteads counted; there may well have been others not included in the lists. Nor does Stolz's list represent *all* communities in the Tyrol, as some students of medieval populations seem to have assumed. These lists do, however, suggest rates of increase that appear reasonable in terms of other estimates and data on the population of the Tyrol. Using these rates as a rough extrapolation backward from 1835, and including Wopfner's estimate for 1650, we get the following figures for the Tyrol as a whole:

POPULATION FIGURES FOR THE TYROL

Year (A.D.)	Total population
1312	240,000
1427	360,000
1650	500,000

Interview Sheet

Name of homestead.
House number.
Owner: date of birth; name of father
and mother.
Wife: date of birth; name of father and
mother.
Genealogies.
Sketch of the life history of each person
appearing in the genealogy.
Inhabitants of the homestead: children.
Inhabitants of the homestead: other.
Economic activities.
History of homestead.
Livestock owned:
    type
    number
    age
    how acquired
    utilization
    livestock products
    disposition
Plots owned:
    plots
    how acquired
    price

where, when, and from whom
    acquired
    use
Types of land owned:
    plowland
    meadowland
    forest
    house and garden
    other
    hectares owned
    location
    type
    how acquired
    utilization
Field produce:
    potatoes
    rye
    wheat
    oats
    buckwheat
    barley
    hay
    straw
Garden produce.
Economic activities by season.

APPENDIX 3

Representative Holdings in Tret and St. Felix

Each case study begins with a brief discussion of the current economic status of an estate or group of closely related estates. This is followed by a detailed, generation by generation, history of inheritance. The genealogies accompanying the case studies include only individuals who lived twenty or more years and thus figured in inheritance strategies. Generations are lettered, beginning with "A" for the most recent generation, "B" for the first ascending generation, etc. Within each generation, birth order is indicated by the number appearing inside each figure. Every individual in a genealogy can be identified by a letter–number combination: for example, A-2, C-4, and D-3.

CASE 1: THE "MILL ESTATE" IN TRET

Although divided in the closing decades of the nineteenth century, this estate is currently one of the largest in the village, consisting of over 10 hectares of land. It is competently managed by B-4, now in his early sixties. He is assisted by his wife and all five of their children, who range in age from early teens to late twenties. In addition to the subsistence income from the land, cash is earned through the daily sale of milk to a dairy and through the operation of a bar and inn. The bar was built in 1965, mainly with capital donated by the three older children, who work outside of the village during the winter. It was located above the village on a mountain meadow with a scenic vista in the hope of attracting tourists who were beginning to find their way into the valley. The venture met expectations, and by 1969 had been expanded into an inn capable of providing tourists with room and board.

GENERATION E

The single individual recorded for this generation probably received the holding from his father, who was also a Tret resident. Inheritance details are unknown.

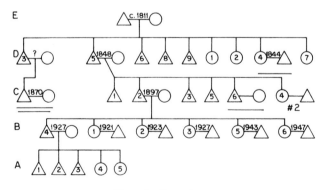

Case 1. Genealogy.

GENERATION D

Holding management: 3 and 5 worked the mill together, and perhaps the land, but each passed ownership of land to his own offspring.

Others: 6 and 7 remained fringe. 1, 2, 8, and 9 left the village, while 4 married a villager in the Val di Non. Inheritance details are unknown.

GENERATION C

LINE I

Continued to operate the mill together with the managers in Line II, but conflict developed between these cousins, including a court case over rights to land. Operation of the mill was eventually carried on by Line II alone. The land associated with Line I was subsequently sold, the owner leaving the village.

LINE II

Holding management: 2 managed the holding alone.

Others: 1 remained fringe until forty-five years old, but died before his father. 3 and 5 migrated to South America. 4 and 6 married in the Val di Non.

Inheritance: D-5 died intestate in 1901, three sons and a daughter surviving him. The daughter 4 was paid off and the three brothers—2, 3, and 6—shared ownership.

1. 2 retained his share and attempted to gain ownership of the shares of his two absent brothers.

2. 3 retained his share, but was willing to turn it over to his brother. Before leaving South America he gave 2 power of attorney so that he could do what he liked with his share. However, 2 never made use of the instrument, and at 2's death, it became worthless.

3. 6 retained his share. Although he married into a prosperous holding in the Val di Non, 6 refused to relinquish ownership, regarding his share as a hedge against some possible but unknown future event that might deprive him of a living on his wife's estate.

GENERATION B

Holding management: 4 manages the holding alone.

Others: All five of the sisters managed to find husbands with holdings in the Val di Non, three of them marrying into one village, two into another.

Inheritance:

1. C-2's share.   C-2 died in 1929, leaving ownership of his share of the holding to his only son, but with the stipulation that each of this son's sisters were to be paid off at one-fifth the value of the estate each. The girls all donated their shares to their brother in 1942.

2. C-6's share.   C-6 died in 1937, his share in ownership going to his only heir, a daughter. This girl had no interest in the share, having also inherited her mother's estate, and in 1942 sold it to B-2, who then became owner of two-thirds of the estate.

3. C-3's share.   C-3 was very successful in Argentina, working first as a laborer, then manufacturing candles in a small factory, and finally buying and operating a large cattle ranch. He married twice, leaving a total of eight offspring when he died, each of them having rights to one-eighth of their father's one-third of the Mill Estate. B-2 exchanged letters with several of these individuals, but lost track of them during World War II.

After the war, he managed to locate them again, through the offices of the Italian consul in Argentina. The heirs he contacted wrote to B-4 that he could have their father's one-third, but the letters would not satisfy the Italian state, which, through its local officer in the Fondo deed's registry office, informed him he would need an affidavit relinquishing rights to the land in his favor, signed by all the living heirs of C-3. There the matter rested until 1961, during which time four of the children of C-3 had died, increasing to twenty the number of heirs to C-3's share of the unsettled estate in Tret.

In 1961, however, B-4 was visited by two of the grandchildren of C-3, both of who had come to Italy to study in a university. One of them, a woman, turned out to be a lawyer, and she said that she would get the necessary signatures when she returned to Argentina. She was as good as her word and collected the signatures of nineteen of the twenty, but the husband of the twentieth was suspicious: suppose the land were of great value? (His wife owns one-sixtieth of the holding!) B-4 has exchanged letters with this man, trying to explain to him how little the land is worth, and offering to pay him anything within reason. None of his offers has been accepted, and in 1965 the matter was still up in the air. However, B-4 had in the meantime found out about the "twenty-year possession law" of 1962, and in 1966 he began proceedings under this law, which gave him ownership of the final share.

CASE 2: THREE BROTHERS' ESTATES IN TRET

All three of these estates are very small, none having as much as a hectare of plowland and meadow, and none holding more than 5 hectares in all (as depicted on the map, estate No. 1 holds 0.83 hectare; No. 2, 0.70; and No. 3, 0.33). None of the managers has an heir apparent living at home who could help ease the labor burden, so the level of operation of all three estates has declined as the men grew older. In 1965, the youngest of the three, B-6 was in his mid-sixties and the other two were both over seventy. Only the youngest brother still keeps cattle, and none tries to do more with his land than raise the few crops necessary to meet his modest needs. In addition, B-2 and B-6 receive small government pensions, while B-5 receives some cash from a widow who lives in his apartment.

GENERATION D

The single individual recorded for this generation probably received the holding from his father, who was also a village resident. Details of the inheritance are unknown.

GENERATION C

Holding management: Following the death of D in 1871, the holding was managed jointly by several of the siblings, principally 4 and 5, for about thirteen years. In 1884, 4 and 5 both married and the holding was divided, with 5 receiving the bulk of the estate. 4 later migrated to Austria and sold his apartment and land, although the apartment was later bought back by 5. 5 continued to manage

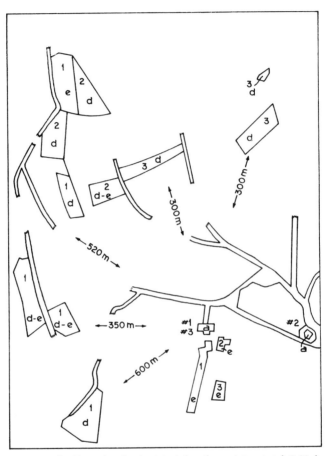

Three estates in Tret: Lands depicted for the estates total 0.83 hectares for estate No. 1; 0.70 for estate No. 2; and 0.33 for estate No. 3. In addition, each estate includes additional mountain meadows on the mountain side above the village.

Map legend:

1	land belonging to Estate No. 1	a	house–barn complex
2	land belonging to Estate No. 2	b	courtyard
3	land belonging to Estate No. 3	c	mill
		d	plowland
		e	meadow

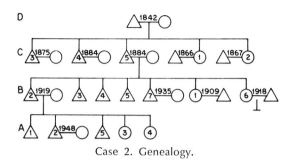

Case 2. Genealogy.

the remainder of the estate and, in 1922, added to it by purchasing a dwarf holding that consisted of an apartment and a few fields.

Others: 3 was fringe for a time, but later operated a combination store and inn, and, since he was also the postman, with a salary from the state, was successful enough to stay in business for some time. Although he married, he had no children, and the business was sold before he died in 1910. Since then the business has had a series of managers but is still in operation. The sisters, 1 and 2, were both married, 1 in a village in the Val di Non, and 2 in Tret.

Inheritance: Each of the five siblings in this generation inherited an equal share of ownership in the holding, but with the stipulation that the sisters be paid off. 4 had the property divided and detached his share: one-fifth of the land and an apartment. The remainder was kept intact by 5, who eventually paid off both of the sisters and also the eldest brother.

GENERATION B

Holding management: The father, C-5 continued as manager until his death in 1934. By this time his four eldest sons were all adult with careers under way, and 7, the youngest son, took over management of most of the estate.

Others: 2 became a stonemason and remained in a fringe relationship to his father's holding, a second apartment being prepared for him in his father's house when he married in 1919. When his father died, he inherited a bit of land, thus becoming a "dwarf" landholder, as well as a mason (this is holding No. 3 on the map).

3 died in a hospital as a soldier in World War I, and 4 migrated to America.

5 remained in a fringe relationship to his natal holding while his father was still manager and worked all his life as a teamster,

staying in the village only occasionally. In 1937 he took in a widow and her children (her husband had been a migrant to the United States and had no land), and she has lived in his apartment ever since, managing the household for him.

The two sisters both married, 1 in Meran and 6 in a village in the Val di Non.

Inheritance: In an unusual move, 6 was "paid off" for her share of the inheritance *before* the death of her father. Her husband had been involved with several other villagers in an enterprise to buy timber from peasants, which they then transported to a lumbermill where it was sold. However, the venture did not prosper, and with bankruptcy inevitable, the husband feared he would lose his estate. To avoid this, a debt to his wife's father, C-5, was fabricated, and the holding turned over to him in payment of the "debt." This estate was then presented by C-5 to his daughter, B-6, as her inheritance, and she and her husband continued to work the holding as before. However, they subsequently left the village for Meran.

At his death, all of C-5's surviving children except 6 (1, 2, 4, 5, and 7) plus his widow, were made heir to his estate in the following way: ownership of his patrimony, the land he had inherited from D, was divided among 1, 2, 4, and 7, but with the stipulation that the migrant son, 4, and the married daughter, 1, were to receive money, thus effectively giving control of the estate to 2 and 7. 2, however, was at the time more interested in his trade as a stonemason. He wanted money instead of land, and an agreement was worked out between 2 and 7 whereby 2 would keep the apartment he had been living in, plus a few parcels of land, and be paid off for the rest of his one-half of the land. Thus, 7 received the other apartment (where his parents had lived) and most of the patrimonial estate, but had to pay off 1 and 4, as well as 2. 7 continues to manage this estate (No. 1 on the map).

The estate that C-5 had purchased in 1922, decidedly smaller than the other, was left to his widow and 5, the teamster, one-half to each. When the widow died in 1941, all her children inherited equal shares of her one-half of the estate, but 5 was left in control and paid the others off (No. 2 on the map).

GENERATION A

Generation B continues in control of the three holdings and no one stands as a successor to management of any of them. B-7 has had no children. B-2 had several children, but the family has been filled with tragedy, all dying untimely deaths except A-2. This individual has left the village, is a full-time stonemason in Meran, and does not intend to return to live in the village. He is, however, the only heir, and will inherit ownership of all three of the holdings.

The only other possible heirs would be the five children of the widow taken in by 5. However, two daughters have married outside the village, and one son has migrated to the United States. One of the other sons works as a carpenter, but visits the village at intervals; the last son was a migrant to the United States, but returned in 1965. He is currently living with 5, and in 1969 had not yet decided what he would do next. However, 5 is not inclined toward leaving his estate to any of them, and intends to leave it to his brother's son, A-2.

CASE 3: THE FOREST WARDEN IN TRET

The main source of income for this holding is derived from the salary its owner earns as a government forester, making it one of the more prosperous in the village. The estate is small, about 5 hectares, and it is not operated at full capacity because the owner tends to neglect it for his forestry work. As a consequence, his wife does most of the work on the land, with occasional help from other villagers. Nevertheless, several cattle are kept, and milk sold to the cooperative dairy in Fondo adds a significant increment to their income.

GENERATION C

The single individual listed for this generation was born in Tret, but no details of his parents are known. He had a dwarf holding, but

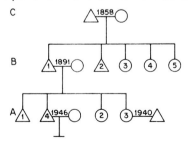

Case 3. Genealogy.

owned no buildings. An apartment and barn were rented from another villager.

GENERATION B

Holding management: 1 managed the holding after his father's death in 1903, except for periods when he was away as a temporary migrant, at which time his wife acted as manager. 1 and 2 both left the village when young, working across the Balkans, European Russia, Siberia, and China. What money they could spare was sent home to their father, who purchased land in their name adjacent to that which he already owned. They returned in 1884, built a house on the land that they had purchased, and had their parents move in with them. They continued in a fringe relation to the holding, the father managing both his own land and land they had purchased. 1 and 2 customarily worked at railroad construction in the lowlands throughout the summer during these years, but 2 returned to Asia, where he died in 1890. 1 migrated to America in 1905, and during the four years that he stayed there sent home money with which his wife purchased more land in the village. After returning home in 1909, he spent most of his time in the village until he died in 1922.

Others: 3 and 4 became nuns, and 5 worked as a housekeeper for a family in Rovereto from 1894 until she died in 1947, a period of fifty-three years.

Inheritance: 1 inherited his father's land and also inherited his brother's, 2, share of the land that they had purchased together. Whether the girls were paid off or not is unknown.

GENERATION A

Holding management: When B-1 died in 1922, all the children were still minors except A-1, and he had migrated to America; therefore, B-1's widow took over management of the holding. She continued as manager until the late 1930s, when 4 took over. In 1934, at age twenty, he had gone into the army and was away on the Ethiopian campaign until 1937. Since returning to the village he has managed the estate, although in 1939 he obtained a job as forester. He receives a salary for this and regards it as his primary occupation. Since his marriage to a Tyrolese girl from St. Felix in 1946 he has left management of the land up to her.

Others: 1 migrated to the United States in 1921 and worked as a shepherd in Western states until his death in 1953. 2 remained a resident of the holding, but died four years after her father. 3 married within Tret.

Inheritance: While all of B-1's offspring, plus his widow, shared equally in ownership of the estate, 2 soon died, leaving the widow, 1, 3, and 4, each owning one-fourth of the estate.

In 1952 4, who was by then manager of the estate, paid off 3 for both her one-fourth of the estate and also for the one-third of the mother's one-quarter share, which 3 would otherwise have inherited when her mother died.

In 1953 1 died; his one-quarter ownership share was divided between his surviving siblings, 3 and 4. 4 then paid off 3 for this share, and since his mother died the same year, he gained full ownership of the holding.

Although childless, 5 and his wife hoped to adopt at least one child, and initiated adoption proceedings in 1965.

CASE 4:   THE "BIG ESTATE" IN ST. FELIX

This estate has resisted division from the time of its establishment, probably in the fourteenth century, until 1965. Although now divided, each of the halves are at least twice as large as any other estate in the village. In both cases the households derive their full income from the land. A-1, in his early forties, is regarded by villagers as a farmer of average skill and ambition. With nine cattle, his herd is over twice the village average, but not up to the capacity of his holding. Even so, his standard of living is above all but a few other households. While still deriving some support from subsistence farming, his main income is from the sale of milk to a lowland dairy. With a newly established household, his sister, A-2, and her husband have few cattle, but were working to increase their herd size and hence their income from the sale of milk.

GENERATION F

The husband in this generation inherited the entire estate intact from his father, but the details of the inheritance are not known.

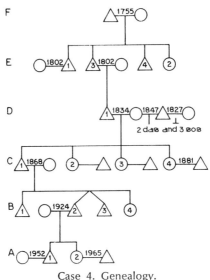

Case 4. Genealogy.

GENERATION E

Holding management: The second son succeeded to management, and ownership, of the entire undivided estate.

Others: The eldest son married into another village estate. The youngest son, 4, and the daughter, 2, left the village.

Inheritance: Details of the inheritance are not known.

GENERATION D

Holding management: The only child to survive to adulthood, a son, inherited the estate. However, he died when his eldest child was but nine years old. His widow was then remarried to a widower who sold his patrimony and came to join her on her late husband's estate, bringing with him two sons by his first marriage. Although C-1, the eldest son of the deceased owner of the estate, inherited the entire property and had succeeded to management at least by 1868 (when he was married), the step-father remained on the holding until his death in 1881.

GENERATION C

Holding management: The only son took over management of the estate, and remained manager until his death in 1922.

Others: The manager's two eldest sisters, 2 and 3, both married estate managers in St. Felix, and the youngest married a landholder in the South Tyrol. His four half-siblings all found spouses with holdings: the two girls and a boy in St. Felix, the other half-brother in the South Tyrol.

GENERATION B

Holding management: The twin brothers and their sister all remained on the holding, one of the twins marrying two years after his father's death in 1922. He was fifty-three years old at the time of his marriage. However, the sister and his wife could not get along, and two separate households were established, the married couple living in one, the sister and the twin brother living in the other. They continued, however, to operate the holding as a single estate.

Others: The eldest brother remained on the holding, but died at the age of twenty-two.

Inheritance: The twin brothers and their sister each inherited one-third ownership of the estate.

GENERATION A

Holding management: The son succeeded to management of the estate following the deaths of his father (in 1947) and uncle (in 1951). Although he had the right to pay off his sister, she persuaded him to divide the estate, and on reaching twenty-one years of age she married a man from Unser Frau without a holding of his own and the estate was divided. The son's wife was an only child and had inherited an estate of her own, her estate and her husband's half-estate now being managed jointly.

CASE 5: THE ESTATE IN THE WOODS

With fewer than 3 hectares of land, all of it of low productivity, only a single cow, and a large family, this manager, B-1, lives on the brink of total disaster.

GENERATION C

This marginal dwarf holding was put together by two brothers around the turn of the century. They were the sixth and eighth of

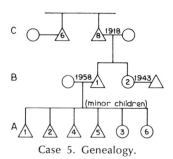

Case 5. Genealogy.

ten children and neither had a patrimony. The land they purchased, although at the same elevation as most village meadows and plow-land, is excessively stony and has no convenient source of water. It had not been cultivated before they bought it and had been left as forest and wasteland. An area of about 5000 square meters was cleared for plowland and meadow, and in 1902 a small house–barn complex was built there. A few years after the house was built, the elder brother, 6, migrated to America. After he returned he bought a holding in Lana and thereafter had nothing more to do with St. Felix. He had been out of contact with the village for over fifty years when he died about 1960. The younger brother thus managed the holding until he died in 1947, although he shared ownership with his absent sibling. The estate was enlarged by the addition of a few parcels of land inherited by his wife.

GENERATION B

C-8's only son succeeded to management of the estate and in-herited ownership of his father's one-half of the holding and his mother's land as well. However, he had to make a cash settlement on his sister. He also had C-6's share to contend with, and in 1949 he bought out this absent uncle's share. In addition, he was able to buy a new, one-half hectare field in 1958. However, the holding will not support his family and in addition to working his land, he must continually seek other sources of income. Some of the problems that he faces as a dwarf-holding manager are:

1. With insufficient plowland he can grow only garden vege-tables and potatoes, no rye or wheat. Therefore, he has to buy all of his bread and do without straw for his animals.

2. With insufficient meadow he cannot grow enough hay to

feed his single milk cow and two heifers, and is forced to buy some hay every year.

3. Because she has insufficient feed, his cow does not give as much milk as she should, not even enough for his children, so he must buy milk daily from other villagers.

4. He does not own a wagon, and even if he did, he could not afford to keep a draft animal or to hire one from a neighbor when needed. He transports hay, potatoes, and wood from field to home in a wheelbarrow, and as a result, everything takes him longer to do than it does anyone else.

In the past he has made and sold rakes, and worked as a herdsman and day laborer in an attempt to make ends meet; in recent years he has given up rake-making because it is too time consuming, and he has not been offered any jobs as a herdsman for some time. He has, thus, been depending on income from occasional employment to supplement the inadequate income from his land, but this has not been sufficient: in 1965, he was forced to sell a meadow in order to keep his family fed. By the following summer this money was gone, and he hoped to go to Austria after the harvest was completed in the fall to find work as a lumberjack (on his later misfortunes, see Chapter IX, pages 230–232).

CASE 6:   FOUR ESTATES IN ST. FELIX

Although the original, undivided estate contained over 7 hectares of plowland and meadow, this land currently supports four independant households (see the accompanying map).

Estate No. 1 is well managed by a man in his mid-sixties. He is assisted in its operation by his wife, who owns the estate, a resident adult son who is the heir apparent, and two adult daughters, both unmarried, who remain in a fringe relationship to the holding. The estate contains nearly 10 hectares of land in all and provides the household with an above average income from subsistence agriculture and the sale of milk. The husband also makes and sells hay rakes and earns a small salary as sexton of the village church.

Estate No. 2 is also well managed by a man in his twenties. He is assisted by his wife, his mother, and a number of siblings in a fringe relationship to the holding, which is about the same size as estate No. 1 in total land. The family is able to derive almost all

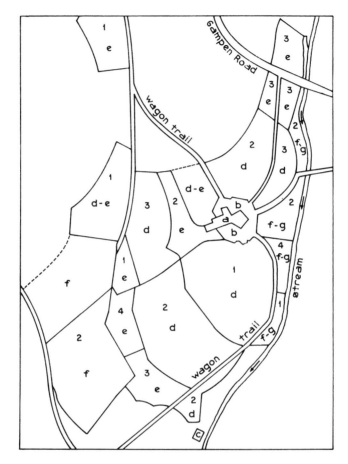

Four estates in St. Felix: Excluding forest and wasteland, the areas of the estates depicted here are 1.80 hectares for No. 1; 3.13 hectares for No. 2; 1.90 hectares for No. 3; and 0.41 hectares for No. 4. In addition, each estate holds additional parcels of meadow and forest on the mountain side above the village.

Map legend:

1	land belonging to Estate No. 1	a	house–barn complex
2	land belonging to Estate No. 2	b	courtyard
3	land belonging to Estate No. 3	c	mill
4	land belonging to Estate No. 4	d	plowland
		e	meadow
		f	forest
		g	wasteland

its support from subsistence agriculture and the sale of milk, although the manager occasionally works for neighbors as well.

Estate No. 3 is small, with fewer than 5 hectares of land. Nevertheless, the manager has been able to live from its output since moving there in 1964. As in the above two cases, income is derived from a combination of subsistence agriculture and the sale of milk. The manager also sometimes works for a brother who owns a sawmill.

Estate No. 4 also has fewer than 5 hectares of productive land. The manager is an excellent carpenter, and through the years he has relied on his craft to augment the income from the land. Now in his sixties, he is semiretired, working his land as best he can and drawing a small pension. While an adult daughter remains at home, his only son has migrated permanently to Germany.

Early history. Although supporting four domestic units at the present time, the original unity of this holding is attested to by the clustering of the four house–barn complexes in a single building mass, and by the distribution of plowland and meadow. Taken as a whole, the parcels belonging to the four current holdings form a solid block of land around the buildings (see map).

The earliest documentary evidence for the holding's existence is from 1423 (Tarneller 1909: 601). During the seventeenth century it was still intact, but by the beginning of the eighteenth century the holding had been divided and was supporting two separate lines with different family names.

The Line I estate remained intact through four generations in the eighteenth and nineteenth century, but during the management of the last Line I manager, a feud developed between the lines. While some of the details of the feud are unknown, it is known that there

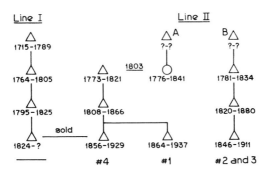

Case 6. Genealogy, Lines I and II.

was a dispute over rights to a parcel of plowland. A court case developed, and when he lost the case, the Line I manager sold the rest of his holding and left the village for good, migrating with his family to Bosnia about 1875.

Line II, meanwhile, had divided, sometime before the last quarter of the eighteenth century. The estate supporting one of these new lines, B, then remained undivided into the twentieth century, passed on every generation from father to son. In the other new line, A, ownership was passed on to a daughter (one of two siblings, both daughters). In 1803, she married and a son eventually took over management of the holding. Following his death the holding was operated by two brothers. Although the feud with Line I probably began with their father, it was during the period of their management that the court case developed, and they purchased the Line I estate when its manager left the village. Thus, during the last twenty-five years of the nineteenth century, this farm was supporting two independent households.

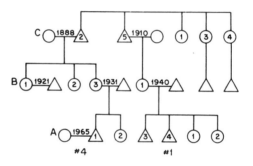

Case 6: Line II-A.

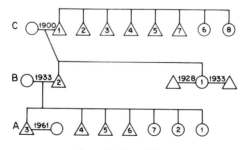

Case 6: Line II-B.

GENERATION C (HOLDING NO. 4)

Holding management: While the two brothers, 2 and 5, operated their patrimony and the estate they had purchased together, the elder brother, 2, was in managerial control of the estate, and in 1888 he married. Twenty-two years later the younger brother married and each took over one of the estates they had earlier combined. As manager of the total estate, the elder brother dictated the division: he took the old Line I holding, leaving his patrimony to the younger brother. However, the elder brother also required that the younger make him a cash settlement.

Others: Each of the three daughters remained in a fringe relationship to the holding throughout their lives, none of them ever marrying. 3 and 4, however, each had a son, both of whom migrated to Germany around 1930.

GENERATION B (HOLDING NO. 4)

The manager of No. 4 died without a will, his three daughters thus inheriting one-third ownership each; in 1931, two years after the death of their father, they worked out a division. The eldest daughter, 1, who had married into another village holding in 1921, received one-third of the land, which has since been worked together with her husband's estate. The rest of the land and the house–barn complex remained intact and was managed by the other two sisters. In 1931, the youngest daughter married, her husband taking over management of the holding. The second daughter eventually left the holding, but retained her ownership share until 1962.

The man who married 3 had been the heir apparent to another village holding, a dwarf estate, but had deserted his patrimony and migrated to Argentina. However, his father died without writing a will, so he and his only sibling, a brother, each inherited one-half of the estate. He returned to the village in 1931, marrying the same year he returned. After joining his wife on her estate, he worked out a division with his brother, keeping about half the land and leaving the rest of the land and the buildings to his brother.

GENERATION A (HOLDING NO. 4)

The daughter, now in her thirties, remains at home with her parents, but the son has migrated to Germany. Trained as a dairyman, he now lives and works in Bavaria. Until his marriage in 1965 he sent a portion of his earnings home each month, but now that he has a family to support, he is no longer able to do this. However, that money was used to pay off his mother's sister, B-2, in 1962. Since he had sent the money, his mother had him recorded as coowner of B-2's share, so that he now has a one-quarter ownership right to the holding.

GENERATIONS C AND B (HOLDING NO. 1)

After his marriage, 5 managed this estate until his death in 1937. His only daughter inherited the estate, and in 1940 she married a man who, as a younger brother, had been disinherited. Prior to his marriage he had spent over twenty years in a fringe relationship to his natal holding, working as a herdsman and a farm laborer outside of the village whenever he could.

GENERATION A (HOLDING NO. 1)

The eldest son remains at home and will succeed to management of the holding. The second brother is learning to be a tailor, and although he visits home occasionally, he does not participate at all in the operation of the estate. Both daughters have permanent jobs outside of the village, one working as a domestic for a family in Cles, the other as a nurse's aid in a hospital in Bozen. Both have remained in a fringe relationship to their natal holding and spend most of the summer and fall at home.

GENERATION C (HOLDINGS NO. 2 AND 3)

While 1 succeeded to management of the holding, all of his siblings remained either core or fringe members of the holding, except the youngest sister, 8, who became a nun. The domestic group consisted of three households: 1 and his wife and children; two brothers and a sister, 3, 4, and 6; and another brother, 7. The first two households consisted of apartments in the main building, the third of a room in a mill some distance from the main house. The three households worked the land jointly, and while 7 managed the mill, which was used not only by neighbors in St. Felix, but by people in neighboring villages as well, all the brothers worked there as needed. 3, 4, and 5 had all gone to South America about 1890 and had sent money home that was used to buy more land, but they returned to the village to stay within a few years.

In 1911 the eldest brother died, his son and heir being only seven years old at the time. However, with so many adults on the holding, there was no trouble keeping it in operation, and 4, who became the new manager of the estate, was appointed guardian of the children. This brother died in 1918, and another took his place. The last of the siblings died by 1937, but 1's son had taken over management in 1933, the year that he married.

The brothers 2 and 5 were both in a fringe relationship to the holding, 2 working as an animal dealer and 5 as a teamster. 2 eventually became completely independent, but 5 remained in the fringe all his life.

GENERATION B

Since each of the Generation C siblings who remained on the holding had held an ownership share, the details of the inheritance are complicated. The end result, however, was that the son inherited one-quarter ownership outright, his sister inherited one-quarter, and they shared ownership of the remaining one-half. The son succeeded to management of the entire holding and was to pay a cash settlement to his sister for her ownership share. Meanwhile, the sister married into another village holding and left home.

GENERATION A

B-2 died in 1956, and his eldest son, 3, succeeded to management of the estate. Until 1961, when 3 married, the holding was operated by the sibling set. However, B-2 had never succeeded in paying off his sister, and in 1964 a problem arose because of this. His sister, B-1, decided that since she had not been paid, she had a right to claim ownership of one-half the estate. Four of her five sons all had an estate or other occupation, and she wanted the estate for the fifth son. The siblings living on the holding did not think her claim legitimate, but did not stop her when she moved her son and his wife into one of the apartments in the building. An informal agreement on land division has been worked out, but the heirs of B-2 had refused to legalize the division by registering it with the state as of 1967. When revisited in 1969, both sides were in the process of investigating the legality of their claims, but it appeared at that time that B-1's son would have to relinquish at least some of this land to A-3.

Interethnic Marriages*

ST. FELIX MEN MARRYING ROMANCE-SPEAKING WOMEN

1. 1723: The woman was member of an Italian family temporarily resident and holding land in St. Felix.
2. 1723: Woman from Malosco. Man owned Erspam.
3. 1729: Woman from Malosco.
4. 1737: Widower on Lochman homestead marries a woman from Brez.
5. 1840: Man, a dependent laborer, marries a woman from Seio. Several children born to this couple in Tret, indicating unstable residence.
6. 1904: Man dependent laborer. Woman from Bresimo.
7. 1910: Man a marginal landowner. Woman from Tret.

* Source: parish archives of St. Felix and Tret.

8. 1914: Man the great-grandson of a Romance family that migrated to St. Felix (C family) at the end of the eighteenth century. Woman the widow of an Italian innkeeper long resident in St. Felix.
9. 1927: Man from C family, marginal landowner. Woman from Silandro/Schlanders.
10. 1938: Man day laborer. Woman from Silandro/Schlanders, sister of woman in item 9.

ROMANCE MEN MARRYING GERMAN-SPEAKING WOMEN

1. 1731: Man member of an Italian family temporarily holding land in St. Felix. Woman from Prugg, St. Felix.
2. 1766: Man from Tret.
3. 1777: Man from Tret, dependent laborer on Graiter.
4. 1778: Man from Tret, section Cavallaia.
5. 1781: Castelfondo man (C family) marries into St. Felix.
6. 1835: Man from Tret, first dependent laborer, then owner of Brunn, marries woman from Laurein.
7. 1836: Man from Fondo, bought Lochman, marries St. Felix woman, resides in St. Felix.
8. 1849: Smith, from Tret, marries St. Felix woman, reside in St. Felix.
9. 1876: Man from Fondo, innkeeper in St. Felix, marries St. Felix woman.
10. 1937: Railroad employee, village unknown, marries St. Felix woman, a waitress.
11. 1937: Landowner in Tret marries a St. Felix woman. She had an illegitimate child.
12. 1944: Worker in road maintenance marries St. Felix woman.
13. 1955: Man from Borgo marries St. Felix woman, becomes landowner.
14. 1958: Fondo man marries St. Felix woman; they reside in Fondo.
15. 1960: Fondo man marries St. Felix woman; they reside in Fondo.

TRET: 366 MARRIAGES ON RECORD

1. 1758: Tret man marries a woman from St. Felix.
2. 1785: Tret man marries a woman from Meran.

3. 1788:   A German speaker, resident of Tret (Kofler), marries Tret woman.

4. 1810:   A German speaker, resident in Tret (Kofler), marries a St. Felix woman.

5. 1838:   A Tret man marries the German-speaking widow of a Romance immigrant to St. Felix.

6. 1860:   Tret man marries a woman from Unser Frau, date imprecise.

7. 1872:   Tret man marries an Innsbruck woman.

8. 1897:   A man from Perdonig, St. Paul, marries a Tret woman.

9. 1912:   Man with a Serbian name, born in Vienna, marries a Tret woman.

10. 1913:   Relative of man in item 9 marries a Tret woman. He is a sawyer.

11. 1925:   Tret man marries a woman from the vicinity of Meran (Burggrafenamt).

12. 1938:   Relative of men in items 9 and 10 marries a Tret woman. He is born in Innsbruck.

13. 1943:   German-speaking railroad worker marries Tret woman.

14. 1945:   Tret man marries woman from Tisens. She has an illegitimate child.

15. 1946:   Tret man marries a woman from Unser Frau.

16. 1947:   Tret man marries a St. Felix woman.

17. 1948:   Tret man marries a woman from Mölten. He is a construction worker in Meran.

18. 1949:   Tret man, returned from the United States, marries a woman from Laurein.

19. 1954:   Tret man marries a woman who has immigrated to Tret from Graun/Curon.

20. 1964:   A German-speaking immigrant to Tret from Graun/Curon marries a Tret woman.

21. 1966:   Tret man marries a woman who has immigrated to Tret from Graun/Curon.

22. 1969:   Tret man marries a woman who has immigrated to Tret from Graun/Curon.

Subtracting two cases of Germans marrying Germans (Cases 3 and 4) and three cases of marriage with the men whose name indicates Serbo-Croatian affiliation, there remain seventeen cases of intermarriage between German and Romance speakers. These constitute 4.6% of the 366 cases of marriage recorded for Tret. Cases of intermarriage between both St. Felix and Tret constitute 4.5%

of all 926 marriages recorded. These comprise thirteen cases of German-speaking men marrying Romance women, and twenty-nine cases of Romance men marrying German-speaking women. Twenty-two of these intermarriages took place since 1900, out of a total of forty-two.

Bibliography

Alatri, Paolo, and Edio Vallini
1960 *La questione dell'Alto Adige.* Florence: Parenti.
Alcock, Antony Evelyn
1970 *The history of the South Tyrol question,* published for the Graduate Institute of International Studies, Geneva. London: Michael Joseph Ltd.
Almond, Gabriel A., and Sidney Verba
1965 *The civic culture: Political attitudes and democracy in five nations, an analytical study.* Boston: Little, Brown.
Altenstetter, Klaus
1968 *Die Siedlungs- und Agrarverhältnisse von Laurein, Proveis und Rumo am Nonsberg,* Schlern-Schriften No. 252. Innsbruck: Wagner.
Anati, Emanuel
1961 *Camonica valley.* New York: Knopf.
Arens, Franz
1903 *Die sozialen Anschaungen der Tiroler in ihren Weisthümern.* Gotha: F. Andreas Perthes.
Arendt, Hannah
1958 *The human condition.* Chicago: University of Chicago Press.
Atz, Karl, and P. Adelgott Schatz
1908 *Die deutschen Seelsorgen in den italienischen Dekanaten und Landesteilen der Diözese Trient.* Bozen: Verlag Alois Auer.
Aufderklamm, Dr. Jakob
1959 Die Rechbretter (Totenbretter) in den deutschen Gemeinden des Nonsbergs, In *Die Deutschgegend am Nonsberg (Südtirol),* edited by R. von Klebelsberg, Schlern-Schriften No. 191. Innsbruck: Wagner. Pp. 109–117.

319

Ausserer, Carl
  1899   Der Adel des Nonsbergs, *Jahrbuch der heraldischen Gesellschaft "Adler,"*
         New Series, Vol. 9. Pp. 13–252.
Bar, Le Duc de (Otto Habsburg)
  1935   *Coutumes et droits successoraux de la classe paysanne et l'indivision
         des propriétés rurales en Autriche.* Louvain: École des Sciences Politiques
         et Sociales, Université de Louvain.
Barbieri, Giuseppe
  1958   I mestieri degli emigranti 'e alcune caratteristiche correnti di emigrazione
         della montagna italiana, *Studi geografici publicati in onore del Prof.
         Renato Biasutti,* Supplemento al volume 65 della Rivista Geografica
         Italiana. Florence: La Nuova Italia. Pp. 45–65.
Barth, Fredrik, Editor
  1969   *Ethnic groups and boundaries: The social organization of cultural differ-
         ences.* Boston: Little, Brown.
Bass, Alfred
  1921   *Deutsche Sprachinseln in Südtirol und Oberitalien.* Leipzig: Verlag
         Deutsche Zukunft.
Battisti, Carlo
  1909   Die Nonsberger Mundart. Lautlehre, *Sitzungsbericht der Philosophisch-
         Historischen Klasse der Kaiserlichen Akademie der Wissenschaften* **160:**
         Abhandlung 3.
  1910   Zur Lautlehre der Nonsberger Mundart, *Revue de dialectologie romane*
         2: 345–372.
  1936   La posizione dialettale del Trentino, *Atti del III Congresso Nazionale di
         Arti e Tradizioni Popolari, Roma,* 63–75.
  1922   *Studi di storia linguistica e nazionale del Trentino.* Florence: Le Monnier.
  1929   Anaunia, *Enciclopedia Italiana* **3:** 139.
  1931   *Popoli e lingue nell' Alto Adige; Studi sulla latinità altoatesina.* Florence:
         Bemporad e figlio.
  1959   *Sostrati e parastrati nell'Italia preistorica.* Florence: Le Monnier.
Battisti, Cesare
  1917   *Il Trentino; cenni geografici, storici, economici con un' appendice su
         l'Alto Adige,* 2nd ed. Novara: Istituto geografico de Agostini.
Benedikt, Heinrich
  1958   *Die wirtschaftliche Entwicklung in der Franz-Joseph Zeit, Wiener His-
         torische Studien,* Vol. 4. Vienna: Verlag Herold.
Bertagnolli, Giuseppina
  1958
  –59   *Studio dell'ambiente di Tret.* Quaderno di Geografia, Scuola elementare,
         Classe III. Handwritten manuscript in possession of elementary school,
         Tret.
Bertagnolli, Lino
  1928   "I tipi di abitazione della Valle di Non," *Bolletino della Reale Società
         Geografica Italiana, Ser.* 6, **5**(7–8): 377–394; **5**(9–10): 495–518.
  1930   *Appunti sull'economia della Valle di Non. Collana di monografie re-
         gionali* ed. no. 1 dalla Società per gli studi trentini, Trento.
  1934   Termini geografici dialettali del comune di Fondo, *Studi Trentini di
         Scienze Storiche* **15:** 193–200.

1935 Bacino del Noce, In *Lo spopolamento montano in Italia III.* Istituto Nazionale di economia agraria. Studi e monografie, No. 16: Le Alpi Trentine, Vol. 1, Part 3, pp. 1–60.

Bidermann, H. J.
1886 *Die Nationalitäten in Tirol und die wechselnden Schicksale ihrer Verbreitung, Forschungen zur deutschen Landes- und Volkskunde* 1(7): 389–475.

Block, Mathilde de
1954 *Südtirol.* Groningen: J. B. Wolters.

Blum, Jerome
1948 *Noble landowners and agriculture in Austria, 1815–48.* Baltimore: Johns Hopkins Press.

Borkenau, Franz
1938 *Austria and after.* London: Faber & Faber.

Bottea, T.
1882 La sollevazione dei rustici nelle valli di Non e di Sole, *Archivio Trentino* 1: 63–77.
1883 Le rivoluzioni delle valli del Nosio negli anni 1407 e 1477, *Archivio Trentino* 2: 3–23.

Braito, Gianantonio
1809 *Affare d' issurrezione in Valdinon.* Alla Direzione di Finanza, Commissionata in Trento, Lavis, July 7th. Handwritten manuscript in possession of the Ferdinandeum, Innsbruck.

Bram, Joseph
1965 Change and choice in ethnic identification, *Transactions of the New York Academy of Sciences Ser. II* 28: 242–248.

Brunner, Otto
1959 *Land und Herrschaft, Grundfragen der territorialen Verfassungsgeschichte Oesterreichs im Mittelalter,* 4th ed. Vienna-Wiesbaden: Rohrer.

Callovini, Carlo Guiseppe
1966 *Guida storica e turistica di Fondo e vicinato.* Fondo: Tipografia Anaune.

Cammett, John M.
1967 *Antonio Gramsci and the origins of Italian communism.* Stanford, California: Stanford University Press.

Ciccolini, Giovanni
1919 L'evangelizzazione della Valle del Noce, *Studi Trentini di Scienze Storiche* 28: 3–15, 131–148, 223–244.

Clark, Grahame
1962 *World prehistory: An outline.* Cambridge, England: Cambridge University Press.

Clough, Shephard
1964 *The economic history of modern Italy.* New York: Columbia University Press.

Cohen, Abner
1969 *Custom and politics in urban Africa: A study of Hausa migrants in Yoruba towns.* Berkeley and Los Angeles: University of California Press.

Cohen, Yehudi
1969 Ends and means in political control: State organization and the punishment of adultery, incest, and the violation of celibacy, *American Anthropologist* 71: 659–687.

Cohn, Norman
　1961　*The pursuit of the millenium*. New York: Harper.
Cole, John W.
　1969a　*Inheritance and social process in the upper Val di Non*. Unpublished doctoral dissertation, University of Michigan, Ann Arbor.
　1969b　Economic alternatives in the Upper Nonsberg, *Anthropological Quarterly* **42**: 186–213.
　1970　Le successioni ereditarie e le loro consequenze sociali, *Sociologia: Rivista di Studi Sociali*, New Ser. **4**: 133–158.
　1971　*Estate inheritance in the Italian Alps*, Research Reports No. 10. Amherst: Department of Anthropology, University of Massachusetts.
　1972　Cultural adaptation in the eastern Alps, *Anthropological Quarterly* **45**(No. 3): 158–176.
Correll, Ernst
　1927　Anabaptism in the Tyrol: A review and discussion, *The Mennonite Quarterly Review* **1**(No. 4): 49–60.
Croce, Benedetto
　1933　Poesia popolare e poesia d'arte, *Scritti di storia letteraria e politica* 28. Bari: Laterza. Pp. 1–64.
Cucchetti, Gino
　1939　*Storia di Trento dalle origini al fascismo*. Palermo: Palumbo.
Dami, Aldo
　1960　Les Rhétoromanches, *Le Globe, Bulletin de la Société de Géographie de Genève* **100**: 25–68.
Dedic, Paul
　1939　The social background of the Austrian Anabaptists, *Mennonite Quarterly Review* **13**: 5–20.
Demarchi, Franco
　1968　*Sociologia di una regione alpina, studi e ricerche*, Vol. 7. Bologna: Società Editrice Il Mulino.
Deutsch, Karl W.
　1966　*Nationalism and social communication*. Cambridge, Massachusetts: MIT Press.
Doob, Leonard W.
　1962　South Tyrol: An introduction to the psychological syndrome of nationalism, *Public Opinion Quarterly* **26**: 172–184.
Dopsch, Alfons
　1930　*Die ältere Wirtschafts- und Sozialgeschichte der Bauern in den Alpenländern Oesterreichs*, Ser. A, Vol. 9. Oslo: Instituttet for Sammenlignende Kulturforskning.
Dörrenhaus, Fritz
　1959　*Wo der Norden dem Süden begegnet: Südtirol, ein geographischer Vergleich*. Bozen: Athesia.
Dörrer, Anton
　1947　Hochreligion und Volksglaube. Der Tiroler Herz-Jesu-Bund (1796 bis 1946) volkskundlich gesehen. In *Volkskundliches aus Oesterreich und Südtirol*. Hermann Wopfner zum 70. Geburtstag dargebracht. Oesterreichische Volkskultur: Forschungen zur Volkskunde, Vol. 1. Vienna: Oesterreichischer Bundesverlag für Unterricht, Wissenschaft und Kunst. Pp. 70–100.

Egger, Rudolf
1965 Der Alpenraum im Zeitalter des Ueberganges von der Antike zum Mittelalter. In *Reichenau-Vorträge, 1961–62 (1965)*, Vol. 10. Konstanz-Stuttgart: Jan Thorbecke Verlag. Pp. 15–28.

Ehrenberg, Richard
1963 *Capital and finance in the age of the Renaissance: A study of the Fuggers and their connections.* New York: Kelley.

Eisenstadt, S. N.
1963 *The political systems of empires.* New York: Free Press.

Ennen, Edith
1953 *Frühgeschichte der Europäischen Stadt.* Bonn: Ludwig Röhrscheid Verlag.

Evans-Pritchard, E. E.
1940 *The Nuer.* Oxford: Clarendon.

Fenz, Emanuel, G.
1967 *South Tyrol 1919–1939: A study in assimilation.* Unpublished doctoral dissertation, University of Colorado, Boulder.

Fernwall, R.
1950 Warum haben die Südtiroler optiert?, In *Südtiroler Almanach.* Innsbruck: Wagner'sche Universitäts-Buchhandlung. Pp. 52–60.

Fiebiger, Herbert
1959 *Bevölkerung und Wirtschaft Südtirols.* Bergisch Gladbach: Heider Verlag.

Fortes, Meyer
1958 Introduction. In *The Developmental Cycle in Domestic Groups,* Cambridge Papers in Social Anthropology No. 1, edited by Jack Goody. Cambridge, England: Cambridge University Press. Pp. 1–14.

Franch, Leone
1953 *La Valle di Non.* Trento: Artigianelli.

Franz, Günther
1965 *Der Deutsche Bauernkrieg,* 7th ed. Darmstadt: Wissenschaftliche Buchgesellschaft.

Freeman, Susan Tax
1970 *Neighbors: The social contract in a Castilian hamlet.* Chicago: University of Chicago Press.

Gammillscheg, Ernst
1923 Review of Carlo Battisti, "Studi di storia linguistica e nazionale del Trentino" and "Questioni linguistiche ladine," *Zeitschrift für romanische Philologie* **43**: 247–252.

Gasser, P. Vinzenz
1901 Geschichte des ehemaligen Klosters, der Wallfahrt und Pfarre Senale—Unsere Liebe Frau im Walde am Nonsberg, *Ferdinandeum Zeitschrift Ser.* 3 (No. 35): 80–126.

1906 Das Urbarbuch des Pfarrwidums in U. L. Frau im Walde—Senale—vom Jahre 1524, *Ferdinandeum Zeitschrift Ser.* 3 **50**: 504–520.

Gatterer, Claus
1968 *Im Kampf um Rom: Bürger, Minderheiten und Autonomie in Italien.* Vienna: Europa Verlag.

Geizkofler, Lukas
1961 Kindheit in Sterzing. In *Die Brennerstrasse: Deutscher Schicksalsweg von Innsbruck nach Bozen,* edited by Südtiroler Kulturinstitut. Bozen: Südtiroler Kulturinstitut. Pp. 127–139.

Gellner, Ernest
1969　*Saints of the Atlas.* Chicago: University of Chicago Press.

Germani, Gino
1972　Mass society, social class, and the emergence of fascism. In *Masses in Latin America,* edited by Irving L. Horowitz. London and New York: Oxford University Press. Pp. 577–600.

Giulani, C.
1884
–93　Documenti per la storia della guerra rustica nel Trentino, *Archivio Trentino* 1884, **3**: 95–116; 1887, **6**: 67 passim; 1889, **8**: 1–50; 1889, **9**: 1–48; 1893, **11**: 123–210.

Giusti, Ugo
1943　*Caratteristiche Ambientali Italiane Agrarie, Sociale, Demografiche 1815–1942,* No. 27. Rome: Istituto Nazionale di Economia Agraria, Studi e Monografie.

Grandi, L.
1898　*La guerra rustica nel Trentino nel 1525.* Cles: Visintainer.

Gray, Robert F.
1964　Introduction, In *The family estate in Africa: Studies in the role of property in family structure and lineage continuity,* edited by Robert F. Gray and P. H. Gulliver. London: Routledge & Kegan Paul. Pp. 1–34.

Greenfield, Kent Roberts
1967　The Italian nationality problem of the Austrian empire, *Austrian History Yearbook* 3(Part 2): 491–526.

Günther, Adolf
1930　*Die alpenländische Gesellschaft.* Jena: Fischer.
1951　Tirols Gewerbe an der Schwelle des 19. Jahrhunderts. In *Tiroler Wirtschaft in Vergangenheit und Gegenwart,* edited by H. Gerhardinger, F. Egert, and F. Huter, Schlern-Schriften 77. Innsbruck: Universitätsverlag Wagner. Pp. 281–304.

Habakkuk, H. J.
1955　Family structure and economic change in nineteenth century Europe, *Journal of Economic History* **15**: 1–12.

Hantsch, H.
1953　Die Nationalitätenfrage im alten Oesterreich. Das Problem der konstruktiven Reichsgestaltung, *Wiener Historische Studien,* Vol. 1. Vienna: Verlag Herold.

Hassinger, Herbert
1964a　Der Aussenhandel der Habsburgermonarchie in der zweiten Hälfte des 18. Jahrhunderts. In *Die wirtschaftliche Situation in Deutschland und Oesterreich um die Wende vom 18. zum 19. Jahrhundert, Forschungen zur Sozial- und Wirtschaftsgeschichte,* edited by Friedrich Lütge, Vol. 16. Stuttgart: Gustav Fischer Verlag. Pp. 61–98.
1964b　Der Stand der Manufakturen in den deutschen Erbländern der Habsburgermonarchie am Ende des 18. Jahrhunderts. In *Die wirtschaftliche Situation in Deutschland und Oesterreich um die Wende vom 18. zum 19. Jahrhundert, Forschungen zur Sozial- und Wirtschaftsgeschichte,* edited by Friedrich Lütge, Vol. 16. Stuttgart: Gustav Fischer Verlag. Pp. 110–176.

Haushofer, Albrecht
1928 *Pass-Staaten in den Alpen.* Berlin-Grunewald: Kurt Vowinckel Verlag.
Heer, Friedrich
1966 *The intellectual history of Europe.* London: Weidenfeld & Nicolson.
Hermes, Peter
1952 *Die Südtiroler Autonomie, Volkrecht und Politik,* Vol. 3. Frankfurt: Alfred Metzner Verlag.
Heuberger, Richard
1956 Die Einfügung des Ostalpenraums ins Römerreich. In *Festschrift für Dr. Raimund von Klebelsberg zu Thumburg,* Schlern-Schriften 150. Innsbruck: Wagner. Pp. 105–119.
Higham, C. F.
1967 Stock raising as a cultural factor in prehistoric Europe, *The Prehistoric Society, Proceedings of the Historical Society* 33(No. 6): 99–103.
Hirn, Josef
1909 *Tirols Erhebung im Jahre 1809.* Innsbruck: Heinrich Schwick.
Hitler, Adolf
1926 *Die Südtiroler Frage und das deutsche Bündnisproblem.* Munich: F. Eher.
Huter, Franz
1964 Zur Frage der Gemeindebildung in Tirol. In *Die Anfänge der Landgemeinde und Ihre Wesen* (Konstanzer Arbeitskreis für Mittelalterliche Geschichte), Vol. 1. Konstanz-Stuttgart: Jan Thorbecke Verlag. Pp. 223–235.
1965a *Südtirol: Eine Frage des europäischen Gewissens,* edited by Franz Huber. Munich: R. Oldenbourg Verlag.
1965b Option und Umsiedlung. In *Südtirol: Eine Frage des europäischen Gewissens,* edited by Franz Huter. Munich: R. Oldenbourg. Pp. 338–361.
1965c Wege der politischen Raumbildung im mittleren Alpenstöck. In *Reichenau-Vorträge, 1961–62* (1965), *Die Alpen in der europäischen Geschichte des Mittelalters, Vorträge und Forschungen des Konstanzer Arbeitskreis für mittelalterichte Geschichte,* Vol. 10. Konstanz-Stuttgart: Jan Thorbecke Verlag. Pp. 245–260.
Inama, Vigilio
1900 Il castello e la giurisdizione di Castelfondo nella Valle di Non, *Archivio Trentino* year 15, Fasc. 2: 135–217.
1905 *Storia delle Valli di Non e di Sole nel Trentino dalle origini fino al secolo XVI.* Trento: Ditta Editrice Giovanni Zippel.
1931 *Fondo e la sua storia.* Rovereto: Tipografia Mercurio.
Jäger, Albert
1881 *Die Entstehung und Ausbildung der Socialen Stände und ihrer Rechtsverhältnisse in Tirol von der Völkerwanderung bis zum XV. Jahrhundert.* Innsbruck: Wagner'sche Universitäts-Buchhandlung. 2 vols.
Jászi, Oscar
1961 *The dissolution of the Habsburg monarchy.* Chicago: University of Chicago Press.
Kann, Robert A.
1957 *The Habsburg empire: A study in integration and disintegration.* New York: Praeger.
1967 *The problem of restoration: A study in comparative political history.* Berkeley and Los Angeles: University of California Press.

Klebelsberg, Raimund von, Editor (1959). See von Klebelsberg, Raimund, Editor.

Kolossa, Tibor
1961   Beiträge zur Verteilung und Zusammensetzung des Agrarproletariats in der Oesterreichisch-Ungarischen Monarchie. In *Studien zur Geschichte der Oesterreichisch-Ungarischen Monarchie.* Budapest: Studia Historica Academiae Scientiarum Hungaricae. Pp. 239–265.

Kramer, Hans
1954   *Die Italiener unter der österreichisch-ungarischen Monarchie, Wiener Historische Studien,* Vol. 2. Vienna: Verlag Herold.
1955a   Der Partito Popolare im Trentino vor 1914, *Südtirol: Land europäischer Bewährung, Kanonikus Michael Gamper zum 70. Geburtstag,* Schlern-Schriften No. 140. Innsbruck: Wagner. Pp. 157–168.
1955b   Fürstbischof Dr. Cölestin Endrici von Trient. *Innsbrucker Beiträge zur Kulturwissenschaft* 4: 153–162.
1956   Fürstbischof Dr. Cölestin Endrici von Trient während des ersten Weltkrieges. Nach neu gefundenen Akten, *Mitteilungen des Oesterreichischen Staatsarchivs* 9: 484–527.

Křižek, Jurij
1965   Beitrag zur Geschichte der Entstehung und des Einflusses des Finanzkapitals in der Habsburger Monarchie in den Jahren 1900–1914, in Konferenz der Geschichtswissenschaftler, *Die Frage des Finanzkapitals in der Oesterreichisch-Ungarischen Monarchie 1900–1918.* Bucharest: Verlag der Akademie der Sozialistischen Republik Rumänien. Pp. 5–53.

Ladurner, P. Justinian
1867   Beiträge zur Geschichte des grossen Bauernrebells, 1525 (Nons- und Sulzberg), *Archiv für Geschichte und Altertumskunde Tirols* 4: 85–179.
1868   Regesten aus tirolischen Urkunden, *Archiv für Geschichte und Altertumskunde Tirols* 5: 321–352.

La Palombara, Joseph
1964   *Interest groups in Italian politics.* Princeton, New Jersey: Princeton University Press.

Latour, Conrad
1962   *Südtirol und die Achse Berlin-Rom, 1938–1945, Schriftenreihe der Vierteljahrshefte für Zeitgeschichte,* No. 5. Stuttgart: Deutsche Verlags-Anstalt.

Leach, Edmund R.
1961   *Pul Eliya: A study of land tenure and kinship.* Cambridge, England: Cambridge University Press.

Leidlmair, Adolf
1958   *Bevölkerung und Wirtschaft in Südtirol, Tiroler Wirtschaftsstudien* No. 6. Innsbruck: Universitätsverlag Wagner.
1965a   Bevölkerung und Wirtschaft, 1919–1945. In *Südtirol: Eine Frage des europäischen Gewissens,* edited by Franz Huter. Munich: R. Oldenbourg. Pp. 362–382.
1965b   Bevölkerung und Wirtschaft seit 1945. In *Südtirol: Eine Frage des europäischen Gewissens,* edited by Franz Huter. Munich: R. Oldenbourg. Pp. 560–580.

Le Roy Ladurie, Emanuel
1971   *Times of feast, times of famine: A history of climate since the year 1000.* Garden City, New York: Doubleday.

Lévi-Strauss, Claude
1952   *Race and history.* Paris: Unesco.
Lewis, Oscar
1970   *Anthropological essays.* New York: Random House.
Liber pertinens ad Ecclesiam Curatialem S: S. Felicis Et Antonij Senalli in quo notantur reditus et expensa singulis Annis Secundum ordinem mansum Cujus Curatia
1762   Archive of the Parish of St. Felix.
Lichtenberger, Elisabeth
1965   Das Bergbauernproblem in den Oesterreichischen Alpen: Perioden und Typen der Entsiedlung, *Erdkunde* **19:** 39–57.
Loss, Giuseppe
1877   *L'Anaunia. Saggio di Geologia delle alpi Tridentine.* Trento: Giovanni Seiser.
Lutz, Wilhelm
1968   Das Bild der bäuerlichen Siedlung in Tirol, Beiträge zur Genese der Siedlungs- und Agrarlandschaft in Europa, *Erdkundliches Wissen,* Heft 18; *Geographische Zeitschrift, Beihefte.* Wiesbaden: Franz Steiner Verlag. Pp. 103–111.
Macartney, C. A.
1968   *The Habsburg empire 1790–1918.* London: Weidenfeld & Nicolson.
Maček, Josef
1960   *Tyrolská selská válka a Michal Gaismair.* Prague: Československà Akademie Věd.
Mack Smith, Denis
1969   *Italy: A modern history.* Ann Arbor: University of Michigan Press.
Maffei, Jacop' Antonio
1805   *Periodi istorici e topografia delle valli di Non e Sole nel Tirolo meridionale.* Rovereto: Marchesani.
Maine, Henry Sumner
1963   *Ancient law.* Boston: Beacon Press.
Makkai, Lázló
1960   Die Entstehung der gesellschaftlichen Basis des Absolutismus in den Ländern der österreichischen Habsburger, *Études historiques,* Vol. 1. Budapest: Academie Hongroise des Sciences. Pp. 627–667.
Marchetti, Tullio
1934   *Luci nel buio, Trentino Sconosciuto, 1872–1915.* Trento: Scotoni.
1960   *Ventotto anni nel servizio informazioni militare.* Trento: Museo del Risorgimento.
Mayer, Theodor
1959   Über die Freiheit der Bauern in Tirol und in der Schweizer Eidgenossenschaft. In *Beiträge zur Geschichtlichen Landeskunde Tirols, Festschrift für Universitätsprofessor Franz Huter,* edited by Ernst Troger and Georg Zwanowetz, Schlern-Schriften No. 207. Innsbruck: Universitätsverlag Wagner. Pp. 231–240.
Mayr, Michael
1907   Welschtirol in seiner geschichtlichen Entwicklung, *Zeitschrift des Deutschen und Oesterreichischen Alpenvereins* **38:** 63–92.

1917   *Der italienische Irredentismus. Sein Entstehen und seine Entwicklung vornehmlich in Tirol.* Innsbruck: Tyrolia.

Menghin, Alois
1897   Diesseits und jenseits des Gampen. Separatum from *Bote für Tirol und Vorarlberg* 1896, Innsbruck.

Metz, Friedrich
1961   *Land und Leute: Gesammelte Beiträge zur deutschen Landes- und Volksforschung.* Stuttgart: W. Kohlhammer Verlag.

Mignon, Herta
1938   *Ulten und Deutsch-Nonsberg,* M.A. thesis, University of Innsbruck, Innsbruck.

Minghi, Julian V.
1963   Boundary studies and national prejudices: The case of the South Tyrol, *Professional Geographer* **15:** 4–8.

Monteleone, Renato
1960   Problemi e condizioni economiche del Trentino durante l'anessione al Regno Italico (1810–1813), *Studi Storici* Year 1, No. 5: 913–943.
1971   *Il movimento socialista nel Trentino 1894–1914.* Rome: Editori Riuniti.

Moore, Barrington, Jr.
1966   *Social origins of dictatorship and democracy: Lord and peasant in the making of the modern world.* Boston: Beacon Press.

Mussolini, Benito
1911   *Il Trentino Veduto da un Socialista, Quaderni della Voce* No. 8. Florence: Quattrini.

Naroll, Frada,
1960   Child training among Tyrolese peasants, *Anthropological Quarterly* **33:** 106–114.

Naroll, Raoul
1958   German kinship terms, *American Anthropologist* **60:** 750–755.

Naroll, Raoul
(n.d.)   Tyrolese military spirit. Unpublished manuscript.

Naroll, Raoul, and Frada Naroll
1962   Social development of a Tyrolese village, *Anthropological Quarterly* **35:** 103–120.

Nef, John
1941   Silver production in Central Europe 1450–1618, *Journal of Political Economy* **49:** 575–591.

Nibler, Franz
1886   Der deutsche Nonsberg, *Mitteilungen des Deutschen und Oesterreichischen Alpenvereins,* New Ser. **12**(No. 23): 271–274.
1887   *Bilder aus dem Welschen Nonsberg.* Munich: J. Lindauer'sche Buchhandlung.

Paoli, Lino
1960   *Dal Peller al Roén, motivi di vita 'Nonesa.'* Trento: Artigianelli.

Perini, Agosto
1856   *Dizionario geografico statistico del Trentino.* Trento: Tipografia Fratelli Perini.

Pesso, Elio
1954   Irredentismo e opzioni nell' Alto Adige, *Archivio per l'Alto Adige,* year 48: 333–373.

Piccoli, N., Ed.
1955 *A Dieci Anni—la Resistenza e il Trentino, 8 Settembre 1943–4 Maggio 1945*. Trento: Museo del Risorgimento e della lotta per la Libertà.

Picker, Henry
1951 *Hitlers Tischgespräche in Führerhauptquartier 1941–42*. Bonn: Athenäum-Verlag.

Pittioni, Richard
1931 Urzeitliche Almenwirtschaft, *Mitteilungen der Geographischen Gesellschaft Wien* **74**: 108–113.
1954 *Urgeschichte des Oesterreichischen Raumes*. Vienna: Franz Deuticke.
1960 Urgeschichtliches zum Volkstum der Räter, In *Festschrift zu Ehren Richard Heubergers*, edited by Wilhelm Fischer, Schlern-Schriften No. 206. Innsbruck: Wagner. Pp. 111–120.

Pivec, Karl
1961 Italienwege der Mittelalterlichen Kaiser. In *Die Brennerstrasse: Deutscher Schicksalsweg von Innsbruck nach Bozen*, edited by Südtiroler Kulturinstitut. Bozen: Südtiroler Kulturinstitut. Pp. 84–110.

Plant, Fridolin
1885 *Berg-, Burg- und Thalfahrten bei Meran und Bozen*. Meran: Verlag Fridolin Plant.

Polanyi, Karl
1957 *The great transformation*. Boston: Beacon Press.

Pölnitz, Götz von (1949). *See* von Pölnitz, Götz.

Pospisil, Leo
1971 *The anthropology of law: A comparative theory*. New Haven: Yale University Press.

Prati, Angelo
1919 Ricerche di toponomastica Trentina, *Archivio Glottologico Italiano* **18**: 195–275.

Preradovich, Nikolaus von (1955, 1964). *See* von Preradovich, Nikolaus.

Pulgram, Ernst
1958 *The tongues of Italy*. Cambridge, Massachusetts: Harvard University Press.

Radice, Antonio
1960 *La resistenza nel Trentino 1943–1945*. Trento: Museo Trentino del Risorgimento.

Ratzel, Friedrich
1896 Die Alpen inmitten der geschichtlichen Bewegungen, *Zeitschrift des Deutschen und Oesterreichischen Alpenvereins* **27**: 62–88.

Reich, Desiderio
1896 *Nobiliare Trentino*. Trento: Giovanni Seiser.
1898 L'Anaunia Antica, *Archivio Trentino* **14**: 17–28.
1902 *I luogotenenti, assessori e massari delle Valli di Non e di Sole*. Trento: Giovanni Seiser.
1912
–13 I nobili gentili delle Valli di Non e di Sole, *Tridentinum*, year 14: 426–449; year 15: 1–40.

Reichenau-Vorträge 1961–62
1965 *Die Alpen in der Europäischen Geschichte des Mittelalters, Vorträge und Forschungen des Konstanzer Arbeitskreis für mittelalterliche Geschichte*, Vol. 10. Konstanz-Stuttgart: Jan Thorbecke Verlag.

Reishauer, Hermann
1904a Italienische Siedlungsweise im Gebiet der Ostalpen, *Zeitschrift des Deutschen und Oesterreichischen Alpenvereins* **35**: 77–87.
1940b Siedlungen der Deutschen und Italiener im Gebiete der Ostalpen, In *Zu Friedrich Ratzels Gedächtnis*. Leipzig: Seele. Pp. 289–302.
Riedl, F. H., Editor
1955 *Südtirol, Land europäischer Bewährung, Festschrift zum 70. Geburtstag Kanonikus Michael Gampers*, Schlern-Schriften No. 140. Innsbruck, Wagner.
Ritschel, Karl Heinz
1959 *Südtirol: Ein Europäisches Unrecht*. Graz: Verlag Styria.
1966 *Diplomatie um Südtirol*. Stuttgart: Seewald Verlag.
Roberti, Giacomo
1929 Gli antichi rinvenimenti nella Valle di Non fra il Noce e la sponda destra della Novella, *Studi Trentini di Scienze Storiche* year 10: 185–195.
1951 Quadro sinottico dei ricuperi archaeologici germanici nel Trentino, dalla caduta dell'Impero romano d'occidente alla fine del Regno longobardo, *Studi Trentini di Scienze Storiche* year 30: 323–358.
1957 Bricciche di antichità (Val di Non), *Studi Trentini di Scienze Storiche*, year 36: 1–9, 185–191.
Roschmann, Antonio
1740 *Kurtze Beschreibung der Fürstlichen Grafschafft Tyrol*. Innsbruck: Simon Holtzers.
Rusinow, Dennison I.
1961 Terrorism in the South Tyrol, 1961, *Institute of Current World Affairs, Newsletters* DR 38–40.
1969 *Italy's Austrian heritage 1919–1946*. Oxford: Clarendon Press.
Sahlins, Marshall D.
1961 The segmentary lineage: An organization of predatory expansion, *American Anthropologist* **63**: 322–345.
1965a On the ideology and composition of descent groups, *Man* **64**: 104–107.
1965b On the sociology of primitive exchange, In *The Relevance of Models for Social Anthropology*, edited by Michael Banton. A.S.A. Monographs No. 1, London: Tavistock. Pp. 139–236.
1968 *Tribesmen*. Englewood Cliffs, New Jersey: Prentice-Hall.
Sardagna, Giambattista di
1889 *La guerra rustica nel Trentino (1525)*. *Documenti e Note* (R. Deputazione Veneta di Storia Patria. Monumenti Storici), Ser. 4, Misc. No. 6.
Sarti, Roland
1971 *Fascism and the industrial leadership in Italy, 1919–1940: A study in the expansion of private power under Fascism*. Berkeley and Los Angeles: University of California Press.
Schatz, Josef
1955
–56 *Wörterbuch der Tiroler Mundarten*, Schlern-Schriften Nos. 119–120. Innsbruck: Universitätsverlag Wagner.
Scheffel, Paul H.
1914 *Verkehrsgeschichte der Alpen*, Vol. 2: *Das Mittelalter*. Berlin: Dietrich Reimer.
Schindele, St.
1904 *Reste deutschen Volkstums südlich der Alpen*. Cologne: Bachem.

Schmalix, A.
(n.d.) *Schlösser und Adel am Nonsberg.* Brixen and Trento: G. *Moncher.*
Schneider, Jane T.
1965 *Patrons and clients in the Italian political system.* Unpublished doctoral dissertation, University of Michigan, Ann Arbor.
Schneller, Christian
1877 Deutsche und Romanen in Südtirol und Venetien, *Petermanns Geographische Mitteilungen* No. 23: 365–385.
1893 Volksleben der Romanen in Tirol, In *Die Oesterreichisch-Ungarische Monarchie in Wort und Bild,* Vol. 13. Vienna: K. & K. Hof- und Staatsdruckerei. Pp. 299–328.
1899–
1900 *Südtirolische Landschaften.* Innsbruck: Wagner'sche Universitätsbuchhandlung. 2 vols.
Schorger, William D.
1969 The evolution of political forms in a north Moroccan village, *Anthropological Quarterly* **42:** 263–286.
Schreiber, Walter
1948 Die Lage des bäuerlichen Besitzstandes in Südtirol und im Trentino, *Tiroler Heimat* **12:** 93–112.
Schulte, Aloys
1900 *Geschichte des mittelalterlichen Handels und Verkehrs zwischen Westdeutschland und Italien mit Ausschluss von Venedig.* Leipzig: Duncker und Humblot. 2 vols.
Schürr, Friedrich
1965 Die Alpenromanen. In *Reichenau-Vorträge 1961–62* (1965) *Die Alpen in der europäischen Geschichte des Mittelalters, Vorträge und Forschungen des Konstanzer Arbeitskreis für mittelalterliche Geschichte,* Vol. 10. Konstanz-Stuttgart: Jan Thorbecke Verlag. Pp. 201–219.
Sedlmayer, Hans
1938 Die politische Bedeutung des deutschen Barocks (Der 'Reichsstil'). In *Gesamtdeutsche Vergangenheit. Festgabe für Heinrich Ritter von Srbik zum 60. Geburtstag.* Munich: Bruckmann. Pp. 126–140.
S.P.Q.T. (Senatus Popolusque Tridentinus)
1892 *Il Trentino. Saggio Etnografico Storico Politico.* Milan: Fratelli Dumslard.
Sereni, Emilio
1968 *Il capitalismo nelle campagne (1860–1900).* Torino: Einaudi.
Sidaritsch, Marian
1923 Landschaftseinheiten und Lebensräume in den Ostalpen, *Petermanns Geographische Mitteilungen* **69:** 256–261.
Simonsfeld, Henry
1887 *Der Fondaco dei Tedeschi in Venedig und die deutschvenetianischen Handelsbeziehungen.* Stuttgart: Catta. 2 vols.
Sofisti, Leopoldo
1950 *Male di frontera,* Bolzano: Cappelli.
Sölch, J.
1923 Geographische Kräfte im Schicksal Tirols, *Mitteilungen der Geographischen Gesellschaft in Wien* **66:** 13–45.
1931 Raum und Gesellschaft in den Alpen, *Geographische Zeitschrift* **37:** 143–168.

Staffler, Hans
    1932 Erinnerung an die 500 Jährige Einweihung der Wallfahrtskirche Senale:
          Hospiz und Wallfahrt Senale. Bozen: Vogelweider.
Staffler, Johann
    1959 Beiträge zur Geschichte von Unser Frau im Wald und St. Felix. In Die
          Deutschgegend am Nonsberg (Südtirol), edited by Raimund von Klebels-
          berg, Schlern-Schriften No. 191. Innsbruck: Wagner. Pp. 67–80.
Staffler, J. J.
    1839
    –47 Tirol und Vorarlberg. Statistisch und topographisch mit geschichtlichen
          Bemerkungen in zwei Teilen. Innsbruck: 2 vols.
Staffler, Richard
    1959 Die Deutschgegend. Beitrage zur Geschichtskunde der vier deutschen
          Gemeinden auf dem Nonsberg. In Die Deutschgegend am Nonsberg
          (Südtirol), edited by Raimund von Klebelsberg, Schlern-Schriften No. 191.
          Innsbruck: Wagner. Pp. 17–59.
Steinacker, Harold
    1967 Staatswerdung und politische Willensbildung im Alpenraum. Darmstadt:
          Wissenschaftliche Buchgesellschaft.
Steinacker, W.
    1932 Der Begriff der Volkszugehörigkeit und die Praxis der Volkszugehörigkeits-
          bestimmung im altösterreichischen Nationalitätenrecht, Schriften der In-
          stituts für Sozialforschung in den Alpenländern an der Universität Inns-
          bruck, Vol. 9. Innsbruck: Universität Innsbruck.
Stephens, William N.
    1963 The family in cross-cultural perspective. New York: Holt.
St. Felix
    1879 Manuscript in possession of the Commune of St. Felix, Province Bozen,
          Italy.
Stolz, Otto
    1927 Die Ausbreitung des Deutschtums in Südtirol im Lichte der Urkunden.
          Munich: Verlag R. Oldenbourg, 4 vols.
    1930 Die Schwaighöfe in Tirol, Wissenschaftliche Veröffentlichungen des
          Deutschösterreichischen Alpenvereins, Vol. 5. Innsbruck.
    1950 Zur Geschichte der landwirtschaftlichen Dienstboten in Tirol, In Festschrift
          Karl Haff. Innsbruck: Universitätsverlag Wagner. Pp. 185–194.
    1953 Geschichte des Zollwesens, Verkehrs und Handels in Tirol und Vorarlberg
          von den Anfängen bis ins XX. Jahrhundert, Schlern-Schriften No. 108.
          Innsbruck: Universitätsverlag Wagner.
Sturmberger, Hans
    1957 Kaiser Ferdinand II. und das Problem des Absolutismus, Munich: Verlag
          R. Oldenbourg.
Südtiroler Kulturinstitut, Editor
    1961 Die Brennerstrasse: Deutscher Schicksalsweg von Innsbruck nach Bozen.
          Bozen: Südtiroler Kulturinstitut.
Sulzberger, Cyrus L.
    1959 A new Cyprus in the foothills of the Alps, New York Times, July 30th.

Tarneller, Josef
1910
–11 Die Hofnamen im Burggrafenamt und in den angrenzenden Gebieten, *Archiv für Oesterreichische Geschichte* **1910**: 1–308, **1911**: 185–572.
1923 *Tiroler Familiennamen.* Bozen: Tyrolia.
Terray, Emmanuel
1969 *Le Marxisme devant les sociétiés 'primitives.'* Paris: Maspéro.
Tessmann, Friedrich
1957 Tridentum–Anagnis–Anaunia, *Schlern* **31**: 361–372.
Thaler, Josef
1866 *Der Deutsche Anteil des Bistums Trient.* Brixen: Weger.
Toggenburg, Dr. Paul Graf
1925
–26 *Die bäuerlichen Erbteilungssvorschriften und die Beschränkung der Freiteil-
barkeit von Grund und Boden in Tirol bis zum Kriege.* Unpublished doctoral dissertation, University of Innsbruck, Innsbruck.
Tomasini, Giulio
1955 *Le palatali nei dialetti del Trentino. Appunti sopra un'indagine linguistica.*
Rome: Fratelli Bocca.
1960 *Profilo linguistico della Regione Tridentina.* Trento: Arti Grafiche Saturnia.
Ton, Ettore di
1901 Trentino e Tirolo. Note cartografiche e toponomastiche, *Ateneo Veneto*
year 24, **1**: 3–18.
Toniolo, A. R.
1950 Considerazioni geografiche sull'istituto del maso chiuso, *Atti della Acca-
demia nazionale dei Lincei, Ser. 8, Rendiconti, Classe di scienze morali,
storiche e filologiche* **5**: 381–386.
Toscano, Mario
1967 *Storia diplomatica della questione dell' Alto Adige.* Bari: Laterza.
Trafojer, Ambros
1947 *Unsere liebe Frau im Wald.* Bozen: Ferrari-Auer.
Tremel, Ferdinand
1954 *Der Frühkapitalismus in Innerösterreich.* Graz: Leykamverlag.
1969 *Wirtschafts- und Sozialgeschichte Oesterreichs.* Vienna: Franz Deuticke.
Trener, G. B.
1896
–98 Le antiche miniere di Trento, *Società alpinisti Tridentini* **XX** (Annuario):
27–90.
Trento, Provincia di
1958 Assessorato alla Pubblica istruzione e artigianato. Trent: *Il Trentino.*
Trevor-Roper, H. R.
1967 *Religion, Reformation and social change.* New York and London:
Macmillan.
Tyler, J. E.
1930 *The Alpine passes. The Middle Ages, 962–1250.* Oxford: Basil Blackwell.
*Urbario de beni, livelli, capitali, e rendita delle Venerabili Chiese di Santo
Cristofforo, e di Santo Felice in Senal*
1766 Archive of the Parish of St. Felix.
Veiter, Theodor
1965 *Die Italiener in der Oesterreichisch-Ungarischen Monarchie, Oesterreich*

Archiv, Schriftenreihe des Instituts für Oesterreichkunde. Vienna: Verlag von Geschichte und Politik.

Vinogradov, Amal R.
1970 *The Beni Mtir of the Middle Atlas: a study in Moroccan tribalism.* Unpublished doctoral dissertation, University of Michigan, Ann Arbor.

Voltelini, Hans von (1919). See von Voltelini, Hans.

von Klebelsberg, Raimund, Editor
1959 *Die Deutschgegend am Nonsberg (Südtirol)*, Schlern-Schriften No. 191. Innsbruck: Wagner.

von Pölnitz, Götz
1949 *Jakob Fugger: Kaiser, Kirche und Kapital in der oberdeutschen Renaissance.* Tübingen: J. C. B. Mohr.

von Preradovich, Nikolaus
1955 *Die Führungsschichten in Oesterreich und Preussen (1804–1918) mit einem Ausblick bis zum Jahre 1945, Veröffentlichungen des Instituts für europäische Geschichte Mainz*, Vol. 11. Wiesbaden: Franz Steiner Verlag.
1964 Die politisch-militärische Elite in 'Oesterreich,' *Saeculum* **15**: 393–420.

von Voltelini, Hans
1919 Das Welsche Südtirol, *Erläuterungen zum Historischen Atlas der Oesterreichischen Alpenländer*, Akademie der Wissenschaften in Wien, I. Abteilung: Die Landesgerichtskarte, Part 3: Tirol and Vorarlberg. Pp. 95–260.

von Wolkenstein, Marx Sittich
1936
[ca. 1600] *Landesbeschreibung von Südtirol, Schlern-Schriften* No. 34. Innsbruck: Wagner.

von Zingerle, Ignaz
1877
–88 *Schildereien aus Tirol.* Innsbruck: Wagner, 2 vols.

Weigend, Guido G.
1949 *The cultural pattern of South Tyrol.* Chicago: University of Chicago Press.
1950 Effects of boundary changes in the South Tyrol, *Geographical Review* **40**: 364–375.

Weiss, Marianne
1959 *Die rätoromanische Mundart des Hochnonsbergs.* Unpublished doctoral dissertation, University of Innsbruck, Innsbruck.

Widmoser, Eduard
1961 Zu Grosser Höh ein Gewaltig Strass. Zur Verkehrsgeschichte der Brennerstrasse. In *Die Brennerstrasse: Deutscher Schicksalsweg von Innsbruck nach Bozen*, edited by Südtiroler Kulturinstitut. Bozen: Südtiroler Kulturinstitut. Pp. 303–310.

Whiteside, Andrew G.
1962 *Austrian National Socialism before 1918.* The Hague: Martinus Nijhoff.

Wiessner, Hermann
1946 Beiträge zur Geschichte des Dorfes und der Dorfgemeinde in Oesterreich, *Archiv für Vaterländische Geschichte und Topographie, Klagenfurt* **30**.

Williams, George H.
1962 *The Radical Reformation.* Philadelphia: Westminster Press.

Wiskemann, Elisabeth
1966 *The Rome–Berlin Axis: A study of the relations between Hitler and Mussolini.* London: Collins.
Wolf, Eric R.
1953 La formación de la nación, *Ciencias Sociales* 4: 50–62.
1962 Cultural dissonance in the Italian Alps, *Comparative Studies in Society and History* 5: 1–14.
1966 Kinship, friendship, and patron–client relations in complex societies. In *The Social Anthropology of Complex Societies,* edited by Michael Banton. A.S.A. Monographs 4. London: Tavistock Publ.
1969 Society and symbols in Latin Europe and in the Islamic Near East, *Anthropological Quarterly* 42: 287–301.
1970 The inheritance of land among Bavarian and Tyrolese peasants, *Anthropologica, New Series* 12: 99–114.
Wolff, Karl F.
1955 Die Stellung des Ladinischen im Kreise der romanischen Sprache, *Schlern* 29: 240–246.
1956 Nonsberger Schrifttum, *Schlern* 30: 407–408.
Wolkenstein, Marx Sittich von
1936 *See* von Wolkenstein, Marx Sittich.
Wopfner, Hermann
1908 *Die Lage Tirols zu Ausgang des Mittelalters und die Ursachen des Bauernkrieges, Abhandlungen zur mittleren und neueren Geschichte,* Vol. 4. Berlin-Leipzig: Rothschild.
1927 Die völkliche Einheit Tirols, und ihre Entstehung, In *Das Deutschtum im Ausland: Südtirol,* edited by Karl Bell. Dresden: Deutscher Buch- und Kunstverlag, William Berger. Pp. 20–54.
1938 Beiträge zur Bevölkerungsgeschichte der österreichischen Länder, In *Festschrift zum 70. Geburtstag von Alfons Dopsch,* Baden bei Wien: Rohrer. Pp. 191–242.
1951 Zur Geschichte des bäuerlichen Hausgewerbes in Tirol, In *Tiroler Wirtschaft in Vergangenheit und Gegenwart,* edited by Hermann Gerhardinger and Franz Huter, Vol. 1, Schlern-Schriften No. 77. Innsbruck: Wagner. Pp. 203–232.
1951
–60 *Bergbauernbuch* (Vol. I: 1951, Vol. II: 1954, Vol. III: 1960). Innsbruck: Tyrolia. 3 vols.
Zieger, Antonio
1926 *Storia del Trentino e dell'Alto Adige.* Trento: Casa Editrice G. B. Monauni.
1936 *La Lotta del Trentino per l'Unita e per l'indipendenza 1850–1861.* Trento: Collana del Museo Trentino del Risorgimento.
1956 *L'economia industriale del Trentino dalle origini al 1918.* Trento: Arti Grafiche Saturnia.
1960 Andrea Hofer: ricordi delli insurrezione del 1809, *Archivio per l'Alto Adige,* year 54: 101–154.
Zingerle, Ignaz von (1877–88). *See* von Zingerle, Ignaz.

Index